IFMBE Proceedings

Volume 54

The International Federation for Medical and Biological Engineering, IFMBE, is a federation of national and transnational organizations representing internationally the interests of medical and biological engineering and sciences. The IFMBE is a non-profit organization fostering the creation, dissemination and application of medical and biological engineering knowledge and the management of technology for improved health and quality of life. Its activities include participation in the formulation of public policy and the dissemination of information through publications and forums. Within the field of medical, clinical, and biological engineering, IFMBE's aims are to encourage research and the application of knowledge, and to disseminate information and promote collaboration. The objectives of the IFMBE are scientific, technological, literary, and educational.

The IFMBE is a WHO accredited NGO covering the full range of biomedical and clinical engineering, healthcare, healthcare technology and management. It is representing through its 60 member societies some 120.000 professionals involved in the various issues of improved health and health care delivery.

More information about this series at http://www.springer.com/series/7403

Franco Simini · Pedro Bertemes-Filho
Editors

II Latin American Conference on Bioimpedance

2nd CLABIO, Montevideo, September 30 – October 02, 2015

Springer

Editors
Franco Simini
Núcleo de Ingeniería Biomédica
Facultades de Medicina e Ingeniería
Universidad de la República
Uruguay

Pedro Bertemes-Filho
Grupo de Engenharia Biomédica
Centro de Ciências Tecnológicas
Universidade do Estado de Santa Catarina
Brasil

ISSN 1680-0737 ISSN 1433-9277 (electronic)
IFMBE Proceedings
ISBN 978-981-287-926-4 ISBN 978-981-287-928-8 (eBook)
DOI 10.1007/978-981-287-928-8

Library of Congress Control Number: 2015951759

Springer Singapore Heidelberg New York Dordrecht London

The IFMBE Proceedings is an Official Publication of the International Federation for Medical and Biological Engineering (IFMBE)

Printed on acid-free paper

Springer Science+Business Media Singapore Pte Ltd. is part of Springer Science+Business Media (www.springer.com)

Preface

There is something special in the fact that electrically charged ions moving within a living organism can be detected by means of a metallic electronic phenomenon such as an alternating current. ***Bioimpedance*** is the ability of biological tissues to be evaluated from the outside as conductors of electrical currents. The interface between man-made elements and living organisms is a field of research of its own. Imaging based upon reconstruction from surface voltages and currents yields valuable information on the body's structure. Applications of bioimpedance measurements are also useful to estimate the composition of bodies (e.g. lipidic or watery tissues have different impedance behaviours). The combination of imaging techniques such as Magnetic Resonance, Computed Tomography and Ultrasound with Biompedance measurements gives remarkable results in terms of clinically useful instrumentation. Mostly used in human and veterinary medicine and research, Bioimpedance is of growing importance to study plants too, their composition and growth.

CLABIO 2015 is the 2nd Latin American Conference on Bioimpedance. The first, founding meeting was held in Joinville, Brazil in 2012. Its triennial periodicity and a host University located in different Countries is creating a tradition which researchers are actively supporting. Next CLABIO will be held in 2018 in Caldas, Colombia and the following (2021) in one of the candidate cities to be presented in Montevideo.

Pre-congress Courses were held: ***"Bioimpedance Basics"*** by Prof. Orjan Martinsen of the University of Oslo, Norway and ***"Clinical Applications of Bioimpedance"*** by Prof. Carlos González Correa and co workers of the Universidad de Caldas, Colombia. Students received, as part of their special registration, a copy of the 2015 Book by Dr. Martinsen.

Distinguished researchers such as Prof. Carmelo Felice of the Universidad de Tucumán, Argentina and Prof. Rodion Stepanov of Perm Technical University, Russia along with Prof. Pedro Bertemes-Filho of the Universidade do Estado de Santa Catarina (UDESC), Brazil delivered Master Lectures during CLABIO 2015, sharing with the audience their research and insight.

We are greatly thankful to ***Springer SBM***, for the interest and attention for details with which all aspects of the publication of the Proceedings have been taken care of. The following pages are the tangible result of the excelent peer-reviewing work coordinated by Dr. Pedro Bertemes-Filho , as well as the skilled editorial task by Dr. Christoph Baumann and Ms. Heather King. It is also our pleasure to announce the agreement to publish a book by Springer, collecting the most valuable CLABIO 2015 contributions, with the help of Ms. Marta Moldvai. Finally, being an International Federation of Medical and Biological Engineering (IFMBE) sponsored event, the best papers are published in extended versions in the **IFMBE** Journal ***"Medical & Biological Engineering & Computing"***.

We greatly appreciate the contribution of our financial sponsors: Facultad de Ingeniería, Espacio Interdisciplinario and Comisión de Enseñanza, all of the Universidad de la República (Montevideo, Uruguay), IEEE Uruguay and the Pan American Health Care Exchanges (PAHCE), as well as the moral support of our patrons, Facultad de Medicina, Hospital de Clínicas (both in Montevideo, Uruguay), IEEE Argentina, SOCHIB (Chile), SABI (Argentina), SBEB, FITEJ and the Universidade do Estado de Santa Catarina (UDESC) (all three in Brazil). CORAL, the Latin American Coordination of IEEE Societies, held during CLABIO 2015 its Annual Assembly which was then in Uruguay for the first time.

Last but not least, the Uruguay task force has done a tremendous work to put this meeting together and to host our colleagues: thank you all and welcome to all in Montevideo and in the following pages!

Pedro Bertemes-Filho	Prof. Franco Simini
CLABIO 2015 Co-Chair	CLABIO 2015 Chair

CLABIO 2015 Committees

General Chair of the Conference

Franco Simini — Universidad de la República

General co-Chair

Pedro Bertemes-Filho — Universidade do Estado de Santa Catarina (UDESC)

Technical Program Chairs

Eduardo Santos — Universidad de la República
Marcio Nogueira de Souza — Federal University of Rio de Janeiro

Local Arrangement Chairs

Verónica Fernández Damonte — Universidad de la República
Juan Cardelino — Paysandú Universidad de la República

Finance Chairman

Daniel Geido — Universidad de la República

Marketing & Corporate Relations

Rafael Sanguinetti — Charrua Soft) - Global and Uruguay
Diego Suárez Bagnasco — Universidad de la República) - Argentina

Pre-Conference Coordinator

Ørjan Grøttem Martinsen — University of Oslo

Chair IEEE EMBS-Uruguay

Enrique Ferreira — Un. Católica del Uruguay

Coordination of Voluntary Workers

Gabriela Rama — Universidad de la República
Laura Landin — Universidad de la República

Academic Committee

Raul Gonzalez Lima	University of S.Paulo
Carlos Negreira	Universidad de la República
Ørjan Grøttem Martinsen	University of Oslo
Carlos González Correa	Colombia
Carmelo Felice, Tucumán	Argentina
Pedro Bertemes-Filho	Universidade do Estado de Santa Catarina (UDESC)
Franco Simini	Universidad de la República

Invited Speakers and Keynote Speakers

Carmelo Felice	U. of Tucumán
Ørjan Grøttem Martinsen	U. of Oslo
Pedro Bertemes-Filho	Universidade do Estado de Santa Catarina (UDESC)
Carlos González, David Miranda & Clara González	U. Caldas
Rodion Stepanov	Perm, Russia

CLABIO 2015 CONFERENCE TOPICS

Bioimpedance Applications

Bioimpedance Instrumentation

Body and Tissue Composition

Cell Culture and Cell Suspension

Clinical Applications

Electrical Impedance Tomography

Electrode Modelling

Organ and Tissue Impedance

Technological Advances in Bioimpedance

Speakers

Carmelo Felice, U. of Tucumán

Ørjan Grøttem Martinsen, U. of Oslo

Pedro Bertemes-Filho, UDESC

Carlos González, U. Caldas

Rodion Stepanov, Perm, Russia

Sponsors

CLABIO 2015 was possible thanks to the financial help of the

UNIVERSIDAD DE LA REPÚBLICA, Uruguay
through the following:

Facultad de Ingeniería
Comisión Sectorial de Enseñanza
Espacio Interdisciplinario
Núcleo de Ingeniería Biomédica

IEEE Uruguay Chapter

The Pan American Health Care Exchanges (PAHCE)

CLABIO 2015 has also received the nominal support of the following:

UNIVERSIDAD DE LA REPÚBLICA, Uruguay
through the following:
Facultad de Medicina
Hospital de Clínicas

UNIVERSIDADE DO ESTADO DE SANTA CATARINA, Brazil
through the Grupo de Engenharía Biomédica

IEEE Argentina Chapter
SOCHIB Chilean Society of Biomedica Engineering, Chile
SABI Argentinian Society of Biomedical Engineering, Argentina
SBEB Brazilian Society of Biomedical Engineering, Brasil
ISEBI International Society of Electrical Bioimpedance

CORAL Latin American Coordination of Biomedical Engineering Societies has given support and hosts its 2015 Annual Assembly during CLABIO 2015 in Montevideo, Uruguay.

The organizers of CLABIO 2015 are staff and students of the Universidad de la República of Uruguay and the Universidade do Estado de Santa Catarina (UDESC) of Brazil.

Sponsors

Table of Contents

High Precision System for Bioimpedance Measurement

Le Manh Hai[1], Nguyen Thi Thuong[1], and Vu Ngoc Tuan[2]

[1] Institution for Bio-Medical Physics, Ho Chi Minh City, Vietnam
[2] Institution for Electronics, Hanoi, Vietnam

Abstract—**High precision system for bio-impedance analysis includes high-bandwidth low-noise amplifiers and high speed analog-to-digital converter for signal acquiring. The signal should be processed to extract useful information. This paper describes an effective system for bio-impedance processing in frequency domain with bandwidth up to 1MHz. The algorithm has been developed and tested in Institute for Bio-Medical-Physics for high-bandwidth multi-channel system for bio-impedance measurement equipment. Our motivation involves a bio-impedance measurement system for diagnostics of knee inflammation and other diseases.**

Keywords— *bioimpedance measurement; fast algorithm; bioimpedance analysis;*

I. INTRODUCTION

Many bio-impedance measurement systems or devices have built recently. Most of them have the same principle architecture. For example, [6] show a common system that includes 4 parts: 1) Pulse generator, 2) Front-end circuit with electrodes, 3) High speed analog-to-digital converter or oscilloscope and 4) processing unit (Fig.1).

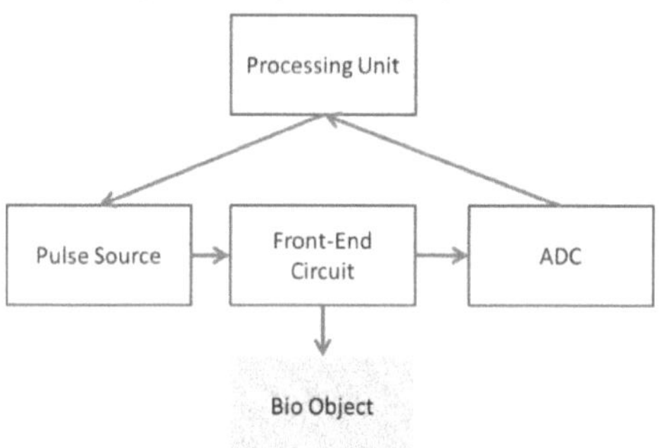

Figure 1: Bio-impedance measurement system

In Fig.1 Pulse source generates periodical rectangle or sinusoidal signal. All parameters of source signal such as pulse form, frequency, amplitude are controlled by Processing unit. Signal is delivered to Bio-object via Front-End Circuit, that provides safe current to Bio-Object and obtains measuring signal via high speed amplifier. Signal from Front-End Circuit is converted from analog to digital by ADC and transferred to Processing Unit for further processing and displaying.

This paper focuses on the algorithm in processing unit. Next section describes more detail of hardware. Section 3 focuses on function of processing algorithm. Some results are represented in section 4.

II. HARDWARE

A. Pulse Source

Although pulse source can be generated by many circuits, Tektronix pulse generator device AFG 2021 has been chosen for its high speed and accuracy, simplicity of control from PC using VISA protocol. AFG 2021 provides rectangle, sinusoidal and triangle pulses with frequency range from 0.01 Hz to 20MHz. (Fig.2)

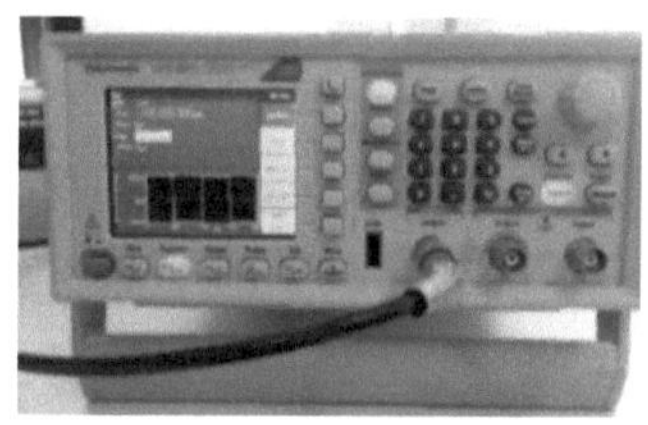

Figure 2: Function generator Tektronix 2021

B. Front-End Circuit with electrodes

Front-end circuit includes constant current source based on Howland circuit (Fig. 3) and high speed fully differential amplifier circuits (Fig. 4)

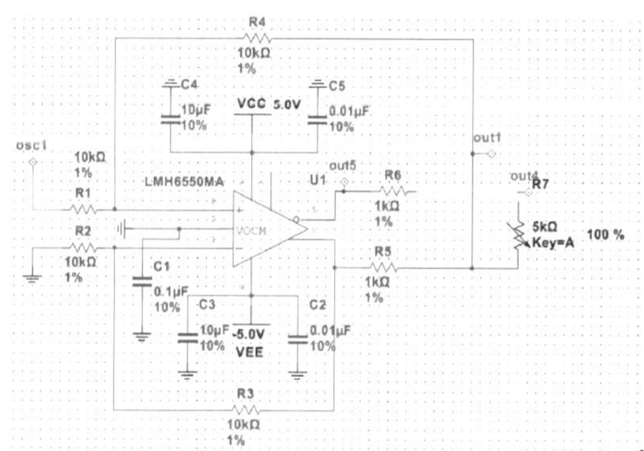

Figure 3: Constant current source.

F. Simini and P. Bertemes-Filho (eds.), *II Latin American Conference on Bioimpedance*,
IFMBE Proceedings 54,
DOI: 10.1007/978-981-287-928-8_1

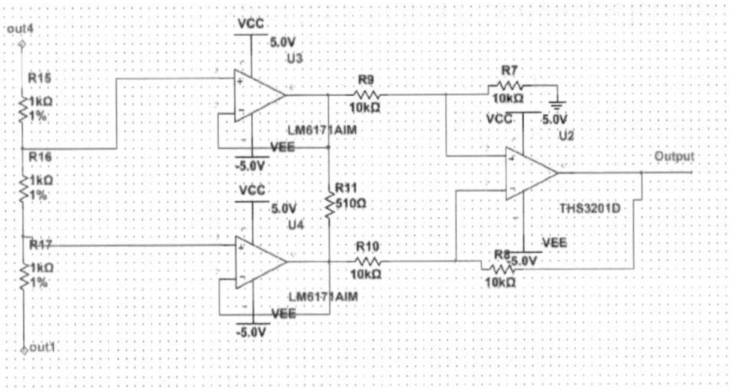

Figure 4: Fully different amplifier circuit

Figure 5: Front-End Circuit

Constant current source provides pulse with same current independent to load. Minimum current is about 1nA, and maximum current is 10 mA. In most measurements amplitude reaches 0.6mA. Front-end circuit wires to bio-object using medical electrodes (Fig. 6)

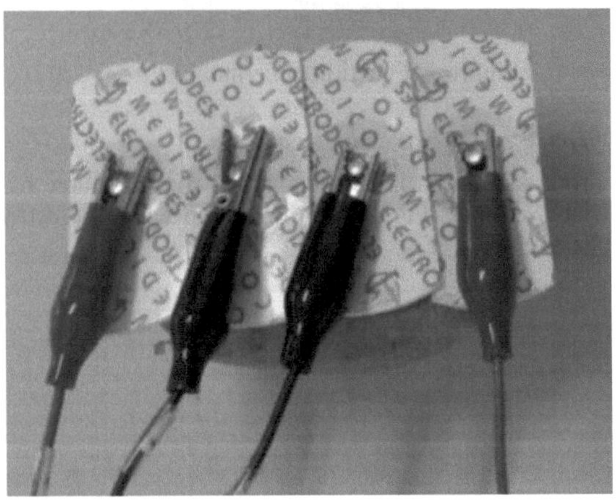

Figure 6: Medical electrode

C. *High speed analog-to-digital converter*

Data acquirement device is high speed Tektronix oscilloscope DPO2022B. This device samples data with bandwidth of 1GSPS. Acquired data is transferred to Processing unit via link with VISA protocol. (Fig.7)

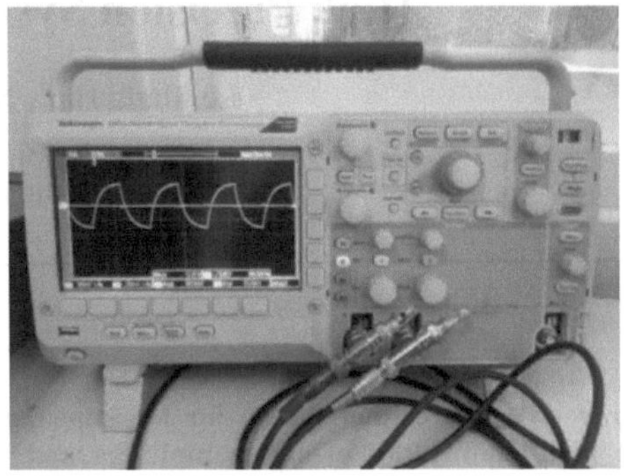

Figure 7: Oscilloscope Tektronix DPO2022B

D. *Processing unit*

Processing unit is high speed PC running Microsoft Windows 7 as operation system and Mathworks MATLAB R2012 as application program.

Fast algorithm for bio-impedance measurement includes six steps as showed in Fig.8

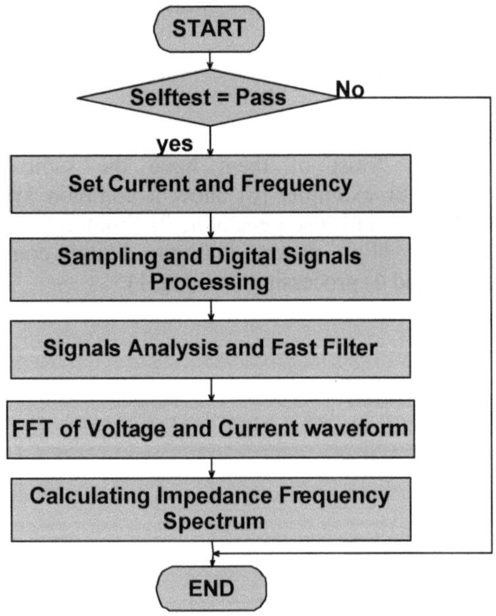

Figure 8: Procedure of data processing

Next section describes the algorithm written in MATLAB.

III. FAST ALGORITHM FOR BIO-IMPEDANCE MEASUREMENT

A. *Calibration step*

In calibration step, a load of 1 KΩ resistor has been used. The system inputs rectangle pulse with amplitude of 100 mV, frequency 1 kHz to the load and calculates system gain and phase delay. Adjust resistor R7, R8 in figure 4 to acquire system gain of 10. Phase delay value is stored for adjustment in sampling step. Calibration step provides system gain about 10 and delay less than 5 ns.

B. Setting up output pulse step

In this step program must issues instruction to pulse generator to output pulse with required frequency and amplitude. Figure 2 shows 1kHz rectangle pulses at the output. Instruction and data has transferred from PC to generator via VISA protocol.

C. Sampling and signal processing step

This is first step of measurement procedure. It requires fast and stable result.

Purpose of the processing module is to calculate bio-impedance in frequency domain. Since ADC device (Tektronix DPO 2022B) gets 10^9 samples per second, system calculate mean value as show in (1)

$$\bar{Z} = \sum_{i}^{n} Zi/n = \overline{Re(Z)} + j\,\overline{Im(Z)} \tag{1}$$

Where Zi is sample value, Re is real part and Im is virtual component, n is amount of samples and equal to 10^9/f, f is frequency of signal. For pulse with frequency of 1kHz, n is 10^6

Accuracy of measurement is (2)

$$\overline{\Delta Re(Z)} = \sum_{i}^{n} \Delta \mathrm{Re}(Z_i)/n \tag{2}$$

$$\Delta \mathrm{Re}(Z_i) = \left|\overline{Re(Z)} - \overline{Re(Z_i)}\right| \tag{3}$$

The algorithm is written in Matlab calculates mean value of samples (4)

```
For i=1:N
        I_temp = I_osc(i);
        U_temp = U_osc(i);
        For j=1:nT
                I_temp = I_osc(i+j*N);
                U_temp = U_osc(i+j*N);
        End
        I_filt = I_temp/nT;
        I_filt = I_temp/nT;
End                                  (4)
```

Signals appear before and after processing step as described in Fig 9.a and 9.b.

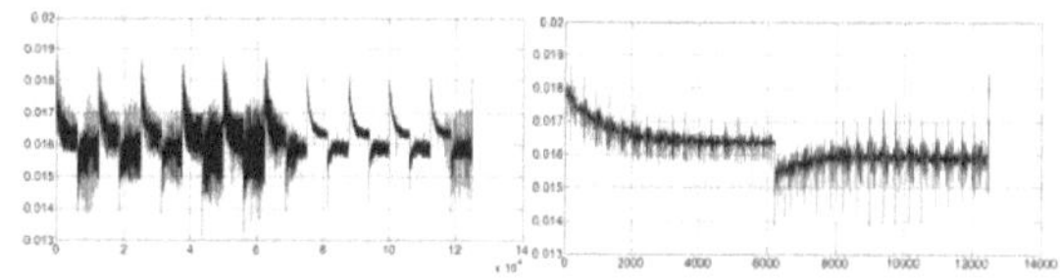

a) *b)*

Figure 9: Signals before processing (a) and after (b)

D. Filter step

This module provides techniques for high frequency filter. In this paper low frequency has been chosen with range from 10^3Hz to 10^5 Hz. This step cleans signal from high frequency noise.

Fig. 10 shows output signal of filter step.

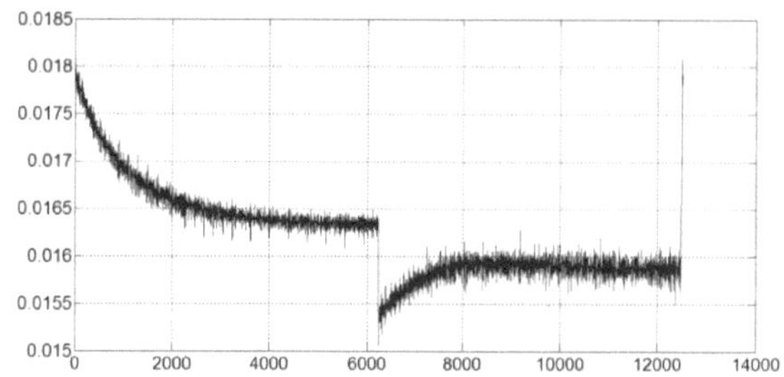

Figure 10: Signals after filter step

E. Fast Fourier transform (FFT)

This step uses FFT build –in function of MATLAB to get bio-impedance spectrum.

Next section will discuss about this result.

IV. EXPERIMENT RESULTS

In experiment, bio-impedance of human arm has been measured. Fig 10 is locations of electrodes in arm of 28 years old man. Where electrodes 1 and 4 wire to current source, electrodes 2 and 3 to front-end circuit. (Fig. 11)

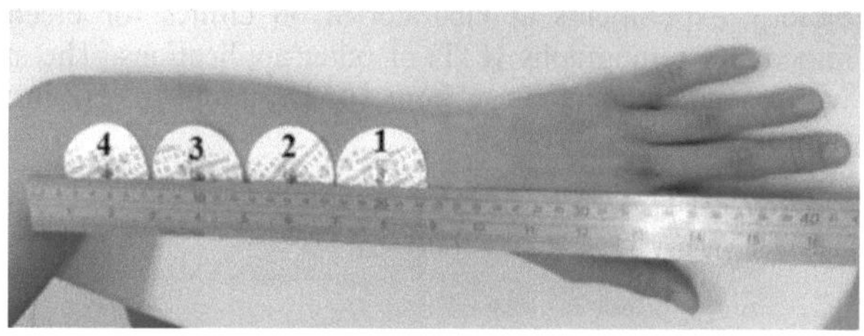

Figure 11: Electrode locations

Periodical rectangle signal with vary frequency from AFG2021 via front-end circuit creates voltage 600mV on electrodes 1 and 4. Voltage and current between electrodes 2 and 3 are measured and processed to get Wessel diagrams (Fig 12). Real part of forearm bio-impedance is about 40 Ω-58 Ω (horizon axis) and imaginary part varies from -1 to -9.6 Ω (vertical axis) in frequency domain from 1 kHz to 100 kHz

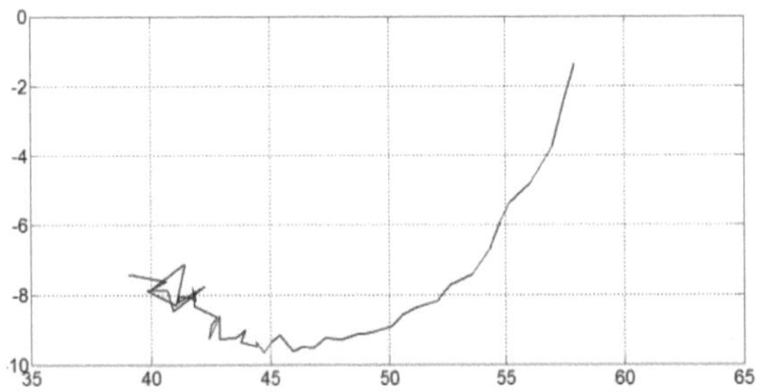

Figure 12: Wessel diagrams for forearm bioimpedance.

Transfer impedance frequency spectrum from forearm is represented in Fig 13.

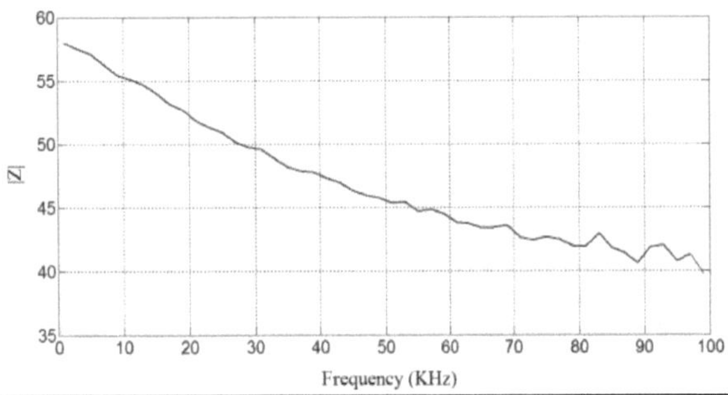

Figure 13: Forearm bio-impedance spectrum: modulus impedance |Z|

These results are homologous with published results in [5] and [6].

V. CONCLUSION

This paper represents a system and fast algorithm for bio-impedance measurement. The system may be used in electro-biology experiments at laboratories on clinics for electrical impedance tomography (EIT) or other applications. The study will extend frequent range from 10^5 Hz up to 10^6 Hz in near future.

CONFLICT OF INTEREST

The authors declare that they have no conflict of interest.

REFERENCES

[1] Jaffrin MY1, Morel H. Body fluid volumes measurement gy impedance: A review of bioimpedance spectroscopy (BIS) and Bioimpedance analysis (BIA) methods. Med Eng Phys. 2008 Dec;30(10):1257-69.

[2] Ling CH, de Craen AJ, Slagboom PE, Gunn DA, Stokkel MP, Westendorp RG, Maier AB. Accuracy of direct segmental multi-frequency bioimpedance analysis in the assessment of total body and segmental body composition in middle-aged adult population. Clin Nutr. 2011 Oct;30(5):610-5

[3] Uwe Pliquett. Bioimpedance: A Review for Food Processing. Food Eng Rev (2010) 2:74–94.

[4] Sami F. Khalil, Mas S. Mohktar and Fatimah Ibrahim. The Theory and Fundamentals of Bioimpedance Analysis in Clinical Status Monitoring and Diagnosis of Diseases. Sensors 2014, 14, 10895-10928;

[5] Uwe Pliquett, Markus Sch¨onfeldt, Andreas Barthel, Dieter Frense, Thomas Nacke and Dieter Beckmann. Front end with offset-free symmetrical current source optimized for time domain impedance spectroscopy. Physiol. Meas. 32 (2011) 927–944

[6] Sverre Grimnes, Orjan G Martinsen. Bioimpedance and Bioelectricity Basics. Third edition. Elsevier 2015

Author: Le Manh Hai
Institute: Institute for Bio Medical Physics
Street: 109A Pasteur, Dist.1
City: Ho Chi Minh City
Country: Vietnam
Email: hailemanh@yahoo.com

Effects of Head Model Inaccuracies on Regional Scalp and Skull Conductivity Estimation Using Real EIT Measurements

M. Fernández-Corazza[1,2,3], S. Turovets[4,5], P. Govyadinov[6], C.H. Muravchik[1,7], and D. Tucker[4,5]

[1] LEICI - Instituto de Electrónica, Control y Procesamiento de Señales, Facultad de Ingeniería, UNLP, Buenos Aires, Argentina
[2] Consejo Nacional de Investigaciones Científicas y Técnicas (CONICET), Argentina
[3] Departamento de Ciencias Básicas, Facultad de Ingeniería, UNLP, Buenos Aires, Argentina
[4] NeuroInformatics Center, University of Oregon, Eugene, OR, USA
[5] Electrical Geodesics Inc. (EGI), Eugene, OR, USA
[6] Electrical and Computer Engineering, University of Huston, TX, USA
[7] Comisión de Investigaciones Científicas de la Provincia de Buenos Aires (CICpBA), Buenos Aires, Argentina

Abstract—**we estimate the scalp and skull conductivities on two healthy adults, based on bounded (or parametric) Electrical Impedance Tomography (bEIT) measurements, and using 62 current injection pairs of a high dense 128 sensor array. We compare the estimates obtained with three different electrode models: pointwise, volumetric, and the Complete Electrode Model (CEM). We also analyze the influence of the skull details and the cerebrospinal fluid (CSF). The estimated scalp (skull) conductivities for these two subjects were ~0.4 and ~0.3 S/m (~0.0045 and ~0.005 S/m), similar for all three electrode models (within 8%). Volumetric and CEM models resulted in a better fit to real data. A model of nested and closed surfaces (no skull holes) resulted in a significant overestimation (~23%) of the skull conductivity. Moreover, neglecting the CSF resulted in an extra 28% overestimation of the skull conductivity. This clearly shows the need of precise head modeling for bEIT.**

Keywords— **bounded electrical impedance tomography, electrode modeling, complete electrode model, parametric estimation, skull conductivity.**

I. Introduction

In EIT, an electric current is applied on the boundary of a conductive object and the resulting potentials are measured by a sensor array on the object's surface. These measurements can be used to estimate the electrical conductivity distribution in the interior of the object. The clinical applications of EIT are numerous [1]. EIT is considered to have a great potential in medical diagnostics as it is a portable, low-cost and non-invasive technique [1]. It can be used to estimate *'in-vivo'* the regional electrical conductivities of the main head tissues in the approach known as parametric or bounded EIT (bEIT) [2-4]. When used to image the internal conductivity distribution of the head, the problem is denoted as EIT imaging or reconstruction [5]. The parametric approach is important for improving EEG source localization [6], targeting in tDCS [7], or EIT reconstruction [5]. The scalp and the skull conductivities have been proven to have a high impact in the accuracy of these applications.

A head model can be obtained from structural magnetic resonance (MR) images, followed by a segmentation (usually from three to seven tissues are differentiated). For each tissue, the conductivity is either assigned or estimated. BEIT is typically considered for the estimation of the scalp and skull conductivities, where most of the injected power is dissipated. Although, a detailed intracranial conductivity distribution can be obtained with Diffusion Tensor Imaging (DTI) [4] or MREIT [8], these techniques are not applicable for the skull conductivity estimation. The scalp and skull can be considered homogeneous and isotropic, or anisotropic with different tangential and radial conductivities [4].

Once the virtual head model is built and the measurements are obtained, the model conductivities are varied to minimize the difference between the measurements and the model predictions. The computation of the scalp potentials is known as the EIT forward problem (FP). It is governed by the Poisson equation and, for complex geometries such as the human head, it can be solved using the Boundary Element Method (BEM) [2], the Finite Element Method (FEM) [4], or the Finite Difference Method (FDM) [9]. FEM and FDM allow inhomogeneities and anisotropies. On the contrary, BEM can only be used in models of nested and closed surfaces of tissue compartments, and assumes homogeneity and isotropy in each layer. The process of fitting the bEIT data and simulations is a non-linear optimization problem. The methods for solving this problem include: the Newton's method (requires up to second order derivative computations) [4], simplex search, and simulated annealing.

Previous bEIT estimates used three layer spherical [10] or BEM [2, 11, 12] models with pointwise or triangular electrodes, and, in [11, 12], only one unknown. A bEIT study using FEM and two unknowns was recently reported [13]. The wide range of the reported skull estimates (form 0.004 to 0.02 S/m) using bEIT and other methods [14, 15] is most likely attributed to the different models employed in those studies. Similarly, the reported scalp conductivity varies from 0.3 to 0.5 S/m, although this is based on a very low number of studies, as reported in [15].

F. Simini and P. Bertemes-Filho (eds.), *II Latin American Conference on Bioimpedance*,
IFMBE Proceedings 54,
DOI: 10.1007/978-981-287-928-8_2

In this work we pursue three specific goals: i) provide new experimental estimates of the scalp and skull conductivity values in detailed seven tissue FEM models; ii) study the influence of the electrode modeling; and iii) study the impact of closed skull three-layer 'BEM-like' models on the accuracy of the bEIT estimates.

II. METHODS

A. bEIT data collection

All research protocols involving human subjects were approved as safe and complying with the ethical standards in the Helsinki Declaration of 1975 by Institutional Review Boards (IRB) at the data collection site (EGI), with informed consent obtained from two subjects (S1, a 46 year old Asian male and S2, a 52 year old Caucasian male) recruited in the studies. Two subjects wearing a 128 channel EGI geodesic net with gelled electrode-to-skin contacts were positioned in a comfortable chair, and 20μA current at the frequency of 27 Hz was administered for 3 seconds in each of 64 distinct electrode pairs. The scalp potentials were acquired and their amplitude extracted using a 256-channel EGI EEG 300 system [3, 16], as well as electrode to scalp impedances. Sensor positions were determined using the Geodesic Photogrammetry System (GPS) [17].

B. Detailed head model construction.

The reference models of soft tissues for S1 and S2 were derived from retrospective T1-weighted MR images obtained with a 3T Allegra scanner (Siemens Healthcare, Erlangen, Germany). Bone structures were derived from retrospective CT scans of S1 and S2 recorded with a GE CT scanner (General Electrics, Fairfield, US). The acquisition matrix was 256x256x256 (voxel size: 1x1x1mm) in both the CT and T1 scans. The T1 MRI images were automatically segmented into seven tissue types (gray matter (GM), white matter (WM), cerebrospinal fluid (CSF), scalp, eyeballs, air, and skull) and the CTs were coregistered to the corresponding MRIs and GPS sensor positions using EGI's segmentation and image processing package, BrainK [18].

Finite element (FE) tetrahedral meshes of ~1.4 million of elements were built from the volumetric segmentations using the iso2mesh package [19] (Fig. 1a). The isotropic and homogeneous conductivity values were assigned to each of the intracranial tissues, based on the typical literature data: 0.2, 0.33, 1.79, and 0.4 S/m for the WM, GM, CSF, and eye balls [15]. The FEM was used for computing the EIT FP. In FEM, the electrical problem is reduced to a linear system of equations $\mathbf{K}(\sigma)\mathbf{v} = \mathbf{f}$, where $\mathbf{K}(\sigma)$ is the 'stiffness' matrix containing the structural and conductivity information, $\mathbf{v}$ is the unknown potential vector at each node of the mesh, and $\mathbf{f}$ is an independent vector that considers the boundary conditions [4]. The linear system was solved using a preconditioned conjugate gradient algorithm [20].

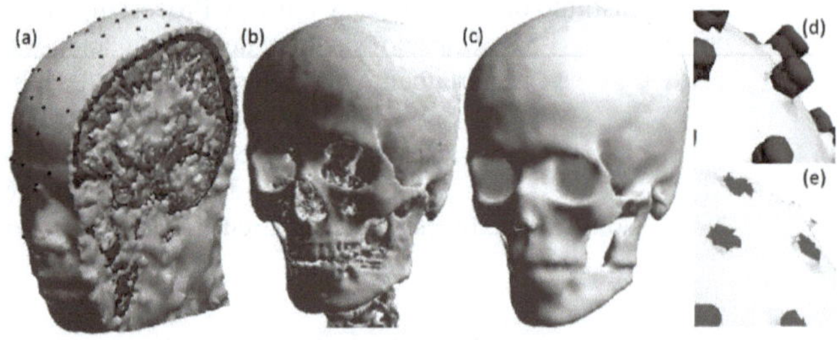

Fig. 1 (a) Computational head model for subject 1 (S1), segmentation, FE mesh, and pointwise electrodes. (b) Skull of the pointwise, volumetric, and CEM models. (c) Skull of the closed skull and closed skull – no CSF models. Details of the (d) volumetric and (e) CEM electrode models on the scalp. Red zones indicate electrode to skin contact surfaces with different boundary conditions in the CEM model.

Within the detailed head models shown in Fig. 1a for S1, we studied three variations of electrode modeling:

Pointwise: each electrode is a node of the tetrahedral mesh. $\mathbf{f}$ is an all zero vector except for two elements: the nodes corresponding to the current injection electrode pair.

Volumetric: the electrodes are modelled as small cylinders of 1 cm height and 5mm radius placed on the scalp. Each cylinder is composed by a thin layer of scalp (2mm), a layer of conductive gel (4mm, 1.5 S/m), and metal (4mm, 500 S/m), as shown in Fig. 1d. The FEM is solved as in the pointwise case, but in this model, current sources and sinks are placed in the metal layers.

CEM: the specific boundary conditions are imposed in the electrode areas (Fig. 1e) in contact with the scalp and, modifying the FEM linear system of equations. The CEM boundary conditions take into account the electrode to skin contact impedances (50KΩ in this study). The details of the CEM can be found elsewhere [5].

For S1, two other model variations were analyzed:

Closed skull: this model is composed by five nested and closed surfaces (scalp, skull, CSF, GM, and WM) and pointwise electrodes. This model is similar to five-layer BEM models. The skull layer is shown in Fig. 1c.

Closed skull – no CSF: this model is the same as previous model except that the conductivity of the CSF layer is set equal to the GM conductivity (0.33 S/m). It is similar to three-layer BEM models where the CSF is not considered.

C. Nonlinear optimization method

For each current injection pair, the estimation process can be formulated as an optimization problem:

$$\widehat{\sigma_{sc}}, \widehat{\sigma_{sk}} = \underset{\sigma_{sc},\sigma_{sk}}{\operatorname{argmin}} \|\mathbf{v_e}(\sigma_{sc}, \sigma_{sk}) - \mathbf{y}\|^2, \quad (1)$$

where σ_{sc}, σ_{sk} are the scalp and skull conductivity values, $\mathbf{v_e}$ is the simulated potential at the electrodes, and $\mathbf{y}$ is the vector with the measurements. Note that (1) is equivalent to the maximum likelihood estimator assuming uncorrelated white Gaussian noise [4]. We used the Newton's method to estimate σ_{sk} and σ_{sc} from (1). This method requires the first and second order derivatives of $\mathbf{v_e}$ with respect to σ_{sk} and σ_{sc}. The computation of these derivatives for the pointwise and volumetric models is detailed in [4]. The formulation of the derivatives for the CEM model were derived specifically for this work in an analogous way.

III. Results

For each head model, the conductivity estimates were calculated for each of 62 bEIT injection pairs separately (two injection pairs were marked as "bad" and discarded). The optimization in (1) was set to stop after 10 iterations of the Newton's method as it was usually stagnated in 7 or 8 iterations. The initial values were set as 0.2 and 0.001 S/m (plus some small random variations) for the scalp and skull conductivities. Lower initial than expected values helps to the convergence of the method [4]. For some particular pairs, and depending on the model, the method did not converge, converged to negative values, or estimated meaningless results. The number of these "outlier" cases was not more than 8 out of 62, and were discarded.

Fig. 2a (Fig. 2b) shows in a box plot the scalp (skull) conductivity estimates for all current injection pairs obtained within each model. Table 1 summarizes the averaged results, and, for each subject, the number of estimations that resulted in a better fit for each model (i.e., a lower norm of the difference between the measurements and the EIT FP computed with the estimates).

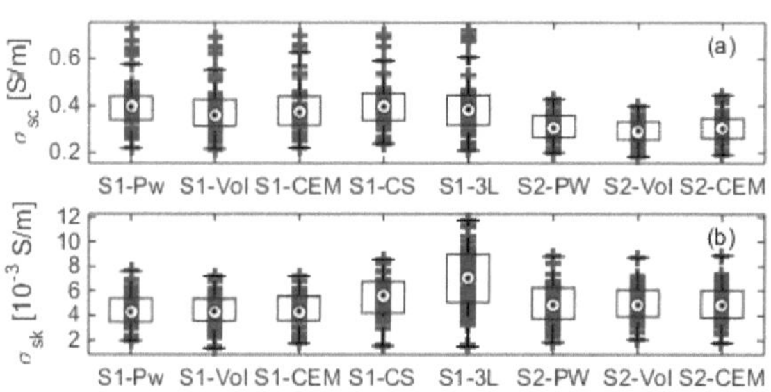

Fig. 2 (a) Scap and (b) skull conductivity estimates for S1 and S2, and for pointwise (Pw), volumetric (Vol), CEM, closed skull (CS), and closed skull – no CSF (3L) models. Red crosses represent individual estimates for each current injection pair, central marks indicate the median, and box edges are the 25th and 75th percentiles.

Table 1 Averaged scalp (SC) and skull (SK) conductivity estimates [S/m], and number of pairs (P) with best fit for both subjects.

Models	S1-SC	S1-SK	S2-SC	S2-SK	S1-P	S2-P
Pointwise	0.401	0.00443	0.310	0.00499	2	13
Volumetric	0.378	0.00439	0.292	0.00497	18	28
CEM	0.389	0.00447	0.305	0.00493	26	13
Closed skull (CS)	0.412	0.00550			2	
CS – no CSF	0.394	0.00705			5	

IV. Discussion

Conductivity estimates: the estimated scalp conductivity is in the middle range of the reported values by similar studies, but the estimated skull conductivity is in the lower limit of the reported range. In addition to the different subject pools, some significant differences might be due to the different models used in each study. In [10], the very first estimations were performed in a quite simple model: a three layer concentric sphere. In [11], with a three layer BEM model and pointwise electrodes, the skull conductivity estimate (~0.01S/m) was similar to the later studies, but scalp was lower (~0.2S/m). Then, in [2], the BEM model used triangles for the electrodes instead of nodes, and the estimated scalp and skull conductivities were ~0.33S/m and ~0.008S/m, respectively. In [12], a similar three layer BEM model was used but the scalp and intracranial conductivities were fixed to 1S/m, thus a one parameter estimation was performed with a result of the scalp/skull conductivity ratio to be ~0.04. Lastly, FEM was used in [13] in a 2D search, with CSF included, resulting in estimates ~0.008S/m for the skull and a rather high value (0.6S/m) for the scalp. We believe that the significant variance of the skull conductivity estimates reported by the literature so far may be attributed to the modeling factors. For this reason, we analyzed the impact of the electrode modeling, the influence of a closed 'BEM-like' skull, and the influence of not considering the CSF layer (similar to a three-layer BEM model).

Differences between estimates can be also explained by inter-subject variability and age [15]. The scalp conductivity difference between S1 and S2 is expected, as S2 is hairless. A hairless scalp is usually dryer and thus, less conductive. On the other hand, the skull thickness in S1 is lower than in S2, implying less spongy bone (more conductive than compact bone) and a less conductive skull.

Influence of the electrode model: the use of pointwise, volumetric, or CEM electrode models resulted in a small difference (<6%) across different estimates of the scalp, and even smaller (<2%) for the skull conductivities. However, volumetric and CEM models showed, in most cases, a better fit between measurements and simulations.

Influence of a closed skull: a closed skull resulted in an approximately high (23%) overestimation of the skull conductivity. This is expected if the skull holes are not modelled, as the estimated skull conductivity has to be higher to compensate for them, i.e., the skull has to be 'more transparent' for the current. This effect might partially explain the higher skull conductivity estimates of previous studies.

Influence of CSF: assuming CSF conductivity equal to GM conductivity led to an extra 28% of skull conductivity overestimation. Adding together these two effects resulted in a skull conductivity estimate of ~0.007 S/m for S1, very close to literature bEIT three-layer BEM estimates.

V. Conclusions

We estimated scalp (skull) conductivities in two subjects to be 0.04 (0.0045) and 0.03 (0.005) correspondingly. Different electrode models showed no significant differences in the estimates, but CEM and volumetric electrode models resulted in a better fit between data and predictions. A closed skull showed a significant overestimation of the skull conductivity value. Neglecting the CSF layer also produces an additional and significant skull conductivity overestimation. It is likely that the combination of both modeling simplifications explain the lower skull conductivity estimated in this study using a detailed FEM model, compared to previous similar studies with simpler BEM models.

The results of this preliminary study should be validated with more subjects, and also more accurate skull models including spatial inhomogeneities should be analyzed.

Acknowledgment

This work was supported by the ANPCyT PICT 2011-11-0909, UNLP 11-I-166, CONICET, CIC-pBA, and the Fundación Bunge & Born, Argentina, the Fulbright Foundation, and the National Institute of Mental Health (Grant R44MH106421), USA. The discussions with the EGI team, in particular Phan Luu, Erik Anderson, and Colin Davey are acknowledged.

Conflict of Interest

Turovets and Tucker are employees of EGI, a manufacturer of dense array EEG systems.

References

1. Bayford, R.H. (2006) Bioimpedance tomography (electrical impedance tomography). Annu. Rev. Biomed. Eng. 8:63-91.
2. Gonçalves, S.I., et al. (2003) In vivo measurement of the brain and skul resistivities using an EIT-based method and realistic models for the head. IEEE Trans. Biomed. Eng. 50(6):754-767.
3. Turovets, S.I., et al. (2008) Conductivity analysis for high-resolution EEG, BMEI 2008. vol. 2, pp 386-393.
4. Fernández-Corazza, M., et al. (2013) Analysis of parametric estimation of head tissue conductivities using Electrical Impedance Tomography. Biomed Signal Proces 8(6):830-837 DOI: 10.1016/j.bspc.2013.08.003.
5. Lionheart, W.R.B., N. Polydordes, and A. Borsic, (2004) Electrical Impedance Tomography: Methods, History and Applications. Inst. Phys., D.S. Holder, Editor, pp 3-64.
6. Vanrumste, B., et al. (2000) Dipole location errors in electroencephalogram source analysis due to volume conductor model errors. Med Biol Eng Comput 38:528-534.
7. Opitz, A., et al. (2015) Determinants of the electric field during transcranial direct current stimulation. NeuroImage 109:140-150 DOI: 10.1016/j.neuroimage.2015.01.033.
8. Sajib, S.Z.K., et al. (2012) Regional absolute conductivity reconstruction using projected current density in MREIT. Physics in Medicine and Biology 57(18):5841 DOI: 10.1088/0031-9155/57/18/5841
9. Turovets, S., et al. (2014) A 3D Finite-Difference BiCG Iterative Solver with the Fourier-Jacobi Preconditioner for the Anisotropic EIT/EEG Forward Problem. Comp. Math. Methods in Medicine:1-12.
10. Eriksen, K.J. (1990) In vivo human head regional conductivity estimation using a three-sphere model, Proc. of the Annual Conf. on Engineering in Medicine and Biology. pp 1494-1495.
11. Oostendorp, T.F., J. Delbeke, and D.F. Stegeman (2000) The conductivity of the human skull: results of in vivo and in vitro measurements. IEEE Trans Biomed Eng 47(11):1487-1492.
12. Clerc, M., et al. (2005) In vivo conductivity estimation with symmetric boundary elements, NFSI2005. vol. 7.
13. Ouypornkochagorn, T., N. Polydorides, and H. McCann. (2014) In Vivo Estimation of the Scalp and Skull Conductivity, EIT2015. pp 10-10.
14. Hoekema, R., et al. (2003) Measurement of the Conductivity of Skull, Temporarily Removed During Epilepsy Surgery. Brain Topogr 16:29-38.
15. Horesh, L. (2006) Some Novel Approaches in Modelling and Image Reconstruction for Multi-Frequency Electrical Impedance Tomography of the Human Brain. PhD Thesis, University College London.
16. Esler, B., et al. (2010) Instrumentation for low frequency EIT studies of the human head and its validation in phantom experiments. J Phys Conf Ser 224(1):012007-012007.
17. Russell, G.S., et al. (2005) Geodesic photogrammetry for localizing sensor positions in dense-array EEG. Clin Neurophysiol 116(5):1130-1140 DOI: 10.1016/j.clinph.2004.12.022.
18. Li, K., A.D. Malony, and D.M. Tucker. (2006) Automatic brain MR image segmentation by relative thresholding and morphological image analysis, VISAPP (1)'06. pp 354-364.
19. Fang, Q. and D.A. Boas. (2009) Tetrahedral Mesh Generation from Volumetric Binary and Gray-scale Images, Proc. of the Sixth IEEE International Conference on Symposium on Biomedical Imaging: From Nano to Macro. Piscataway, NJ, USA pp 1142-1145.
20. Barrett, R., et al., (1994) Templates for the Solution of Linear Systems: Building Blocks for Iterative Methods, 2nd Edition. SIAM, Philadelphia, PA.

Author: Mariano Fernández Corazza
Institute: LEICI, Facultad de Ingeniería, Universidad Nacional de La Plata
Street: Calle 48 y 116
City: La Plata (1900)
Country: Argentina
Email: marianof.corazza@ing.unlp.edu.ar

Electrolytic Extracellular Phantom to Study the Low-Frequency Conductivity of Cervical Neoplasia

Sandra M. Pinto[1,2], Sandra P. Corzo[1], Stelia C. Méndez-Sanchez[1,2], and David A Miranda[1]

[1] Universidad Industrial de Santander, Escuela de Física, CIMBIOS, Bucaramanga, Colombia
[2] Universidad Industrial de Santander, Escuela de Química, GIBIM, Bucaramanga, Colombia

Abstract— **In this work we performed an electrolytic phantom based on an extracellular media containing different ionic species simulating extracellular electrical conductivities to study different cervical neoplastic stage which have been reported in the literature (1). An aqueous electrolytic solution composed by sodium, calcium and potassium salts was selected to emulate the ionic species in the extracellular media. The extracellular conductivities of normal and neoplastic tissues were simulated using different dilutions of extracellular model with the same relation between concentrations of each ionic species. The conductivities values ranged from 1190 µS/cm to 4386 µS/cm. At low frequencies, abnormal tissue shown a conductivity greater than normal tissue. Our results suggests that the extracellular medium of an abnormal tissue can be modeled using an electrolytic solution with a higher concentration of ionic species more than in a normal tissue.**

Keywords— **Ionic concentration, conductivity, neoplasia.**

I. Introduction

Early detection of cancer is a subject of research which requests the study of cellular biochemistry from different points of view. The human body is composed by a high percent of aqueous electrolytic solution divided by membranes in the intra and the extracellular media (2).

The composition of the extracellular fluid includes a high amount of sodium, chloride ions and small quantities of potassium, calcium, magnesium and organic molecules derived from metabolism of proteins.

In order to prevent abrupt changes in cellular pH, cell homeostasis requires a coordinate control in respiratory and renal mechanism as well as in the intracellular and extracellular buffers for maintaining the acid-base equilibrium (3).

When homeostatic mechanisms are altered by a pathological condition the pH of extracellular fluid can decrease or increase (4).

Ionic channels and ion fluxes control can regulate several aspects of tissue homeostasis. During the oncogenic transformation, the function of ion channels can be altered and their deregulation has implications in tumorigenesis and metastasis. Indeed anomalies in ionic channel have been implicated in various aspects of cancer development (5).

Pathological alterations also change the electrical properties of tissue. According to (1), normal intraepithelial tissue exhibit an extracellular electrical conductivity near to 1.2mS/cm and the increase in this value could be associated with the grade of the lesion.

In this paper we present an electrolytic extracellular phantom of normal and abnormal human tissues which can be related with changes in the electrical properties of tissue with consequent variations in the ionic concentration.

II. Materials and Methods

We used NaH_2PO_4; Na_2HPO_4. $7H_2O$ to prepare the buffer solution and KCl, $CaCl_2$ and NaCl salts to establish the ionic concentration in the extracellular medium.

A. Extracellular Medium Preparation

Ionic concentrations of extracellular medium were prepared in accordance to the values reported in literature for plasma and interstitial fluid (3). Sodium chloride (NaCl) (456.3mg), potassium chloride (KCl) (20.0mg) and calcium chloride ($CaCl_2$) (28.0mg) were added at room temperature to the buffer solution of dibasic sodium phosphate heptahydrated (Na_2HPO4. $7H_2O$) (2.84g) and monobasic sodium phosphate (NaH_2PO4) in order to achieve a normal pH of 7.4 for a solution volume of 100 mL to mimic extracellular media composition, our phantom.

B. Preparation of standards and conductivity measurement

Extracellular media of different tissue neoplasia grades were modeled by our phantom diluted into type 1 water. Different dilutions were prepared in order to achieve conductivity values in the range of the extracellular medium conductivities of Normal (NO), low-grade squamous intraepithelial lesion (LSIL), high-grade squamous intraepithelial lesion (HSIL) and Carcinoma (CA) cervical tissues according to (1). Conductivity measures were performed using a 40d HACH HQ portable multi-parameter meter.

F. Simini and P. Bertemes-Filho (eds.), *II Latin American Conference on Bioimpedance*,
IFMBE Proceedings 54,
DOI: 10.1007/978-981-287-928-8_3

III. RESULTS AND DISCUSSION

Table 1 shown the electrolytic concentrations and conductivity values that modeled extracellular media. In previous studies it was found that the changes in conductivity values of cervical tissue can be associated to different levels of neoplastic lesion. Some authors attribute this changes to the alterations suffered by the tissue structure during neoplastic stages (6), however, recent studies suggest that conductivity values also are affected by another factors related to tissue composition (1, 7), e.g. the extracellular conductivity of normal tissue is close to 1.2mS/cm which corresponds to an ionic concentration less than 12.5mmol/L in our phantom (see table 1).

In this sense, values in table 1 suggest that ionic concentration into extracellular media increase according to the degree of malignancy, i.e. electrical conductivity of tissue with LSIL, HSIL and CA can be associated with the ionic concentration of 17.5 mmol/L, 25.0 mmol/L and 50.0 mmol/L respectively.

Tabla 1: Conductivity values and Ionic concentrations of the electrolytic phantom

Molar Concentration [mmol/L]	Conductivity ±0.01 [mS/cm]	Na^{1+} [meq/L]	K^{1+} [meq/L]	Ca^{2+} [meq/L]	Cl^{1-} [meq/L]
200.0	*15.19*	*142.0*	*4.0*	*5.0*	*103.0*
75.0	*6.42*	*53.3*	*1.5*	*1.9*	*38.6*
50.0	*4.58*	*35.5*	*1.0*	*1.3*	*25.8*
37.5	*3.77*	*26.6*	*0.8*	*0. 9*	*19.3*
25.0	*2.81*	*17.8*	*0.5*	*0.6*	*12.9*
17.5	*1.74*	*12.4*	*0.4*	*0.4*	*9.0*
12.5	*1.60*	*8.9*	*0. 3*	*0.3*	*6.4*

Gheorghiu model for cells in suspension (8) suggest that changes into extracellular media can induce changes into intracellular media according to the equation:

$$\sigma_{out} = \sigma_{in}\left(\frac{R_1}{R_1 + d}\right)^2$$

Where σ_{in} and σ_{out} are the charge density at the inner (R_1) and outer ($R_1 + d$) sites of the membrane.

From Gheorghiu model we infer that changes in extracellular ionic concentration are the results of changes in the intracellular ionic concentration. Then, the high concentrations of Na+, Ca2+, K+ y Cl- in extracellular medium observed into HSIL and CA tissues could be associated to deregulations in Ion channels and Transport (ICT) (9). We believe that genetic or epigenetic cellular alterations have been associated to ICT deregulation.

We hypostatize that changes in the ionic concentration affect the cellular metabolism, protein expression and programmed cell dead because firstly, the differences in the ionic concentration depend upon Na, K and Ca cations concentration as well as the proton concentration gradient across the plasmatic membrane and secondly, ionic and pH alterations in the cytoplasm induce damages in the mitochondria such as: membrane potential changes, formation of channels in the outer mitochondrial membrane, release of calcium into the cytoplasm, increased reactive oxygen species, inactivation of oxidative phosphorylation, among others which are related to the induction of some cancer types (10,12,13).

IV. CONCLUSIONS

We conclude that extracellular conductivity changes associated to neoplastic tissue can be related both to modification of ionic concentration of extracellular medium and structural changes in tissue.

Acknowledgment

The financial support of the Universidad Industrial de Santander (UIS), Bucaramanga, Colombia (VIE-5742) and COLCIENCIAS are gratefully acknowledged.

Conflict of Interest

The authors declare that they have no conflict of interest

References

1. Miranda DA, Corzo SP, González-Correa C-A. Cervical cancer detection by electrical impedance in a Colombian setting. J Phys Conf Ser. 2013 Apr 18;434:012056.
2. Walz W. Integrative physiology in the proteomics and post-genomics age. saskatoom: Humana Press; 2007.

3. Hall JE. The Body Fluid and Kidneys. Guyton and Hall Textbook of Medical Physiology. 12th ed. Elsevier Science; 2011.

4. Taylor MM. Electrolytes and Acid–Base Physiology. Integrativephysiology In Theproteomics Andpost-Genomicsage. New Jersey: WOLFGANG WALZ; 2005. p. 27–42.

5. Storm P, Klausen TK, Trulsson M, Ho C S J, Dosnon M, Westergren T, et al. A unifying mechanism for cancer cell death through ion channel activation by HAMLET. PLoS One. 2013 Jan;8(3):e58578.

6. Brown , Brian H, et all. Detection of cervical intraepithelial neoplasia using impedance spectroscopy: a prospective study. junio de 2005, BJOG, Vol. 112, págs. 802-806.

7. Lopamudra, Das, Soumen, Das y Jyotirmoy, Chatterjee. Electrical Bioimpedane Analisis: A New Method in Cervical Cancer Screening. 2015, Journal of Medical Engineering.

8. Gheorghiu E. The resting potential in relation to the equivalent complex permittivity of a spherical cell suspension. Phys Med Biol. 1993 Jul 1;38(7):979–88.

9. Shan Ping Yu LMC and DWC. Ion homeostasis and apoptosis. elsevier Sci. 2001;13:405–11.

10. Prevarskaya N, Skryma R, Shuba Y. Ion channels and the hallmarks of cancer. Trends Mol Med. Elsevier Ltd; 2010 Mar;16(3):107–21.

11. Pedersen SF, Stock C. Ion channels and transporters in cancer: pathophysiology, regulation, and clinical potential. Cancer Res. 2013 Mar 15;73(6):1658–61.

12. Leanza L, Zoratti M, Gulbins E, Szabo I. Mitochondrial ion channels as oncological targets. Oncogene. Nature Publishing Group; 2014 Jan 27;(August 2013):1–13.

13. Szabo I, Zoratti M. Mitochondrial channels: ion fluxes and more. Physiol Rev. 2014 Apr;94(2):519–608.

Bone Electrical Impedance and Tomographic Reconstruction of Fracture Detection: A Review

A.H. Dell'Osa

Instituto de Desarrollo Económico e Innovación, Universidad Nacional de Tierra del Fuego, Ushuaia, Argentina

Abstract—**Electrical Impedance Tomography (EIT) techniques elaborate two dimensions images from average spatial distribution of resistivity within a three-dimensional structure. In the fracture and healing process of long bones, the limb has changes of bioimpedance values. This paper review varius works in bone electrical impedance and tomographic reconstruction, and proposes potential improvements for clinical applications of the current technology to apply in first emergency attention in difficult access areas (p.e.: mountains areas).**

Keywords—**Electrical impedance tomography, osteography, bone fracture healing, transverse resistivity ratio, bioimpedance.**

I. Introduction

Electrical Impedance Tomography (EIT) techniques elaborate two dimension images from average spatial distribution of resistivity within a three-dimensional structure.

When a certain region is crossed by a constant current (frequency range between 10-100kHz and sinusoidal waveform), a complex potential is induced at the boundary, which depends on the regional distribution of resistivity. Sheffield's researchers produce in vivo images from measurements of the boundary voltage gradients followed by backprojection along equipotential lines [1].

This technique is used in clinical applications such as dynamic cardiac imaging, monitoring gastric emptying, pulmonary imaging, monitoring bone healing, and monitoring hyperthermia treatment, in inspite of having a spatial resolution of 5-10% of the diameter of the region [2].

This medical imaging method has some advantages such as being non-invasive, non-hazaradous, high speed-good temporal resolution, sensitive to very small impedance changes and inexpensive [3] and portable equipment.

The aim of this paper is to review the application of EIT in monitoring bone healing as studied by Aberbeen's researchers in the late 80s and early 90s and to propose possible applications and upgrades/improvements for clinical applications with the current technology.

II. Long bone fracture

The diaphyseal region of long bones has a wide range of resistivities due to the presence of different tissues, blood a resistivity of 15MΩ and the bone an average resistivity of 150MΩ. Bones and muscles are anisotropic, being more resistive in their transverse section than their longitudinal section. Figure 1, derived from Geddes and Baker [4], shows the range of resistivities encountered.

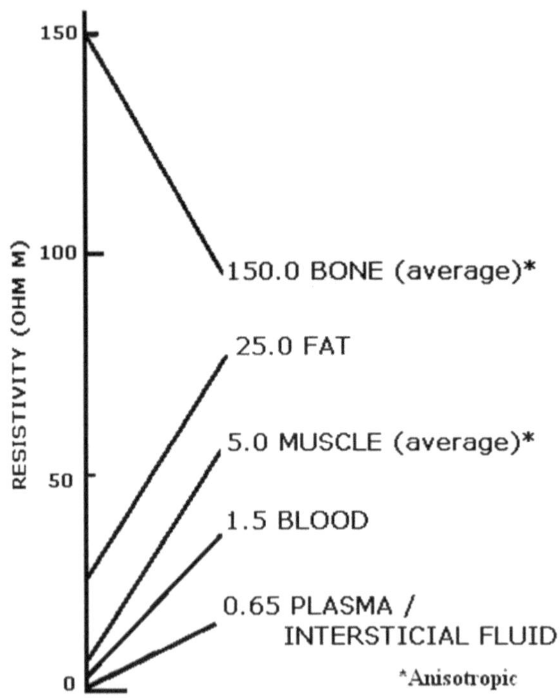

Fig. 1 Bioresistivity values present in human upper arm[2]

Immediately after a fracture, the bone healing process begins and it has three consecutive phases: the inflammatory phase (from 0 to 3 weeks), the reparative phase (from 3 weeks to 6 weeks) and the remodeling phase (more than 6 weeks to years).

F. Simini and P. Bertemes-Filho (eds.), *II Latin American Conference on Bioimpedance*,
IFMBE Proceedings 54,
DOI: 10.1007/978-981-287-928-8_4

Once a diaphyseal fracture occurs, many measurable biopotential changes appear. In the inflammatory phase, while there is a highly conductive region in the plane of the fracture, the resistivity of the limb at the fracture site is very low, being similar to resistivity of blood (1,5MΩ), which is the major component at the fracture hematoma. The causes of resistivity changes are loss of integrity of high resistivity bone, creation of a conducting pathway between the fractured ends of the bone, collection of low resistivity haematoma at and around the fracture site, necrosis of the fractured ends leading to a phase of the inflammatory response with increased vascularity, exudation of low resistivity plasma like oedema fluid in the interstitial space of the surrounding tissue and increased local temperature, as well as varying degrees of damage and swelling of surrounding tissues, depending on the nature of the trauma.

In the reparative phase, the mass of bone producing cells rallies to produce cartilage, callus and bone This proliferation of cells should increase the resistivity in the region to normal bone levels (reference levels). The callus associated with the fracture is prolonged to remodeling phase. The external bridging callus allows the fracture to stabilize and begin the phase of remodeling.

In this phase, the newly formed bone adapts to its function of load bearing, and unwanted bone is removed. Dead bone that may be relevant to the structure of the limb is revitalized by recanalization with Haversian systems or replaced by creeping substitution, gradually recovering the electrical characteristics of the region.

III. THE ABERBEEN IMPEDANCE IMAGING SYSTEM

In the Aberdeen imaging system, 16 electrodes are placed in a split-array equidistantly around the limb (Figure 2). The current is driven through the bottom two electrodes as shown. The reference electrode is located diametrically opposed to the two driving electrodes. The voltage gradient is measured between the reference electrode and each pair of electrodes on either side of the reference electrode. The procedure is then repeated stepwise around the limb. Thus it is possible to build up a composite impedance image of the three-dimensional section of the limb [6][9].

A known current is driven through electrodes 'A' and 'E'. This sets up lines of current flow between the two electrodes. Along these lines, there are regions of equal potential, which run perpendicular to the current lines. A reference electrode, electrode C, is equidistantly located between electrodes 'B' and 'D'. If the current is being passed through an homogeneous medium, then the voltage gradient between 'C' and 'B', and 'C' and 'D' should be identical. However, if a subterranean fault should be present, this would alter the flow of the current and therefore the distribution of the equipotential lines. This will be manifested as a difference between the voltage gradients 'CB' and 'CD' [6].

Sixteen electrodes (made of Ag:AgCl) form the split array. Their size is 12.5mm x 25mm. In the fracture clinic, the electrodes are equidistantly placed on 25mm wide elastic electrode belts.

The induced boundary potentials are measured, demodulated, digitized, averaged and stored on a disk. Meanwhile, measurement and image reconstruction are controlled by a microprocessor.

The image reconstruction is on a 64 x 64 matrix.

The transverse resistivity ratio (TRR) was calculated in each case as the ratio of the average transverse potential gradient of the test limb to that of the contralateral normal limb at equivalent points [5].

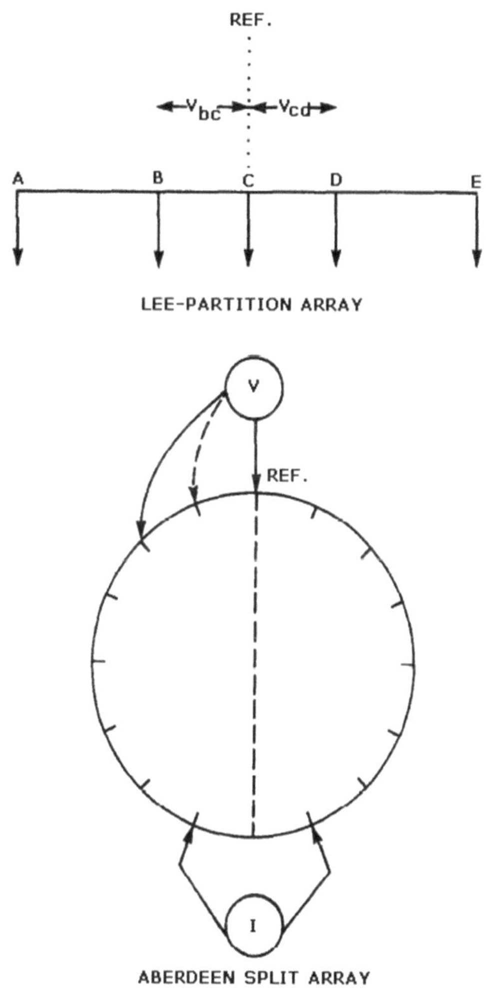

Fig. 2 Split array in Aberbeen impedance imaging system [6]

IV. Review of EIT applied to bone

A research group from Aberdeen University, Scotland, - headed by Dr. Vivek Kulkarni- published a series of articles between 1989 and 1995 that refers to the application of EIT to bone lesions in limbs (humerus [2][5], and tibia-fibula [6]), the improvement of The Aberdeen Impedance Imaging System [7] and a revision of EIT literature [8].

We focus mainly on the articles where the bone fracture of limbs is analyzed using the "Aberdeen Impedance Imaging System".

The obtained results are shown in Figures 3 and 4. Figure 3 shows the values measured in relation with the electrodes placed in the region of interest in the limb, and Figure 3a, the milivolts measured in each pair of electrodes under 3 different situations: healthy bone structure, bone structure 3 weeks after the fracture and bone structure 5 weeks after the fracture.

Figure 3b y 3c show the TRR already described in section III: 'THE ABERDEEN IMPEDANCE IMAGING SYSTEM' based on the measurements obtained from the pair of electrodes used. Figure 3b shows TRR various in very similar situations to 3 curves in healthy humerus, humerus 3 weeks post fracture –contralateral to the healthy arm in the same patient- and humerus 16 weeks post fracture – the same patient's arm now healthy after a fracture. Meanwhile Figure 3c shows the values of a healthy right leg (tibia and fibula), a fracture leg -3½ weeks post trauma-, a successful bone reparation and a bone reconstruction with no union of the segments separated in the fracture.

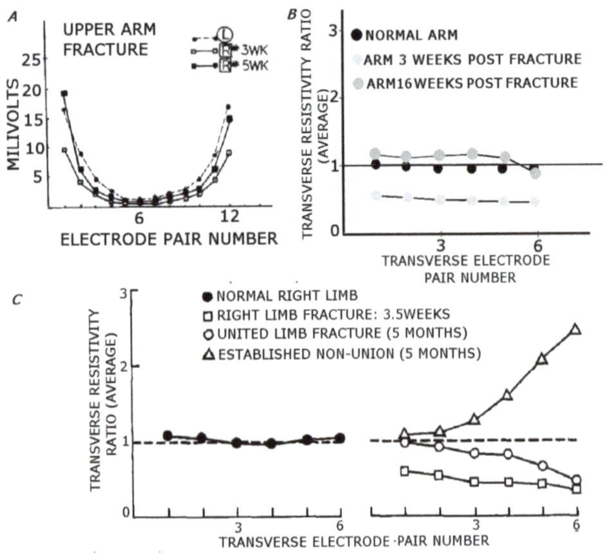

Fig. 3 Impedance measurements in a. humerus[2], b. humerus[5] and c. tibia and fibula[6]

Figure 4 shows the tomographic reconstruction images taking the impedance measurements as a starting point. These images are made of 64 elements high and 64 elements wide. The technology used for the control, processing and digital reconstruction was BBC microprocessor master series [10].

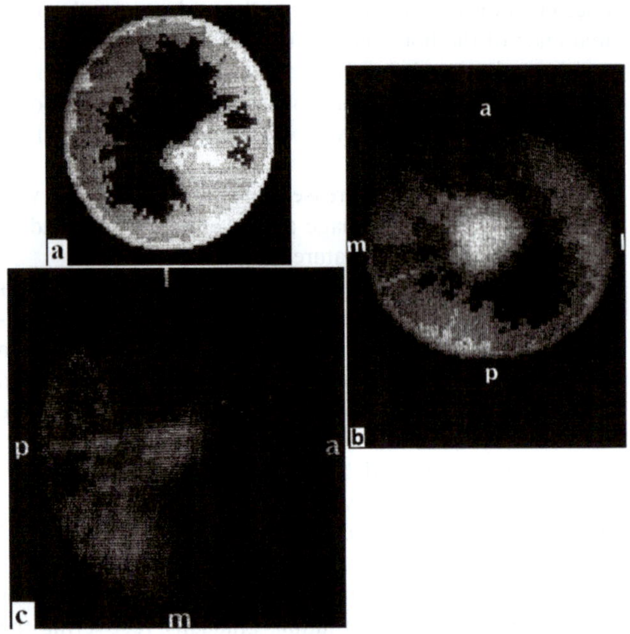

Fig. 4 Tomographic reconstruction of a. right leg, b. right upper limb and c. tibia and fibula [5]

V. Analysis of the results 2 decades later

The results obtained by the work group from Aberdeen University are specific and auspicious.

The measured values (specified in Figure 3) show, at a glance, that there is a variation in the values of bioimpedance in the different situations that can be present in a long bone structure: innate integrity, fracture, integrity after the recuperation and no union after a long recuperation period. These values were measured in the upper limb (humerus) as well as in the lower limb (tibias and fibula).

The tomography axial images reconstructed have no comparable quality and resolution with any of the current images method, either computerized axial tomography, X-Rays or nuclear magnetic resonance, but it is worth noting that the possibility of generating them exist. The difference between bone electrical impedance and others techniques is the mathematical problem that entails the image tomographic reconstruction with some boundaries values.

VI. Possible application and improvements

Una aplicación útil de un método portatil de monitoreo de fractura ósea es para emergencias de primera atención médica en áreas de difícil acceso, como pueden ser zonas de montaña.

The computerized technology used in the works done by the Aberbeen University group, is now obsolete. Taking into account the improvements achieved in the processing of signals and images with the development of microprocessors, Digital Signal Processors and daily use computers, the possibility of getting faster and more exact data is a direct interpolation.

It is also possible to do an analysis starting from the smaller electrodes to generate a boundary's electrode matrix around of one of the human limb that gives a huge quantity of measurable data.

Another possibility is getting measurements using a method different from The Aberdeen Impedance Imaging System, such as the adjacent drive method, the opposite method, the cross method or the trigonometric method [11]. We might also consider the possibility of injecting a current with different parameters.

There is a very complex possibility that would favor the whole EIT technique, and it is finding a new resolution method for the mathematical problem that allows for the reconstruction of the image.

At this time, these improvements proposed are being developed to be evaluated in the near future.

VII. Discussion and Conclusions

The revision of the usefulness of this method in this clinical application appears because of the possibility of getting an ultra portable diagnosis equipment of bone extremity damages (like a notebook plus an additional unit) to places where it is difficult or complex to go with a first emergency medical attention (mountain zone and adverse weather).

It is worth noting that no new contributions for this clinical application were found in the last 20 years, so this may mean that there were no works published.

Nevertheless, the greatest part in the EIT clinical applications is in progress and not many of them would be able to pass from the investigation lab to the production of biomedical equipment. It is for this reason that making an actualization in osteography could offer good results with specific effects and application for the daily use in the medical circle.

Acknowledgment

Natalia Ros and Adriana Barros, for her constant patience and charity.

Conflict of Interest

The authors declare that they have no conflict of interest.

References

1. Barber D C, Brown B H and Seagar A D (1983) Imaging spatial distribution of resistivity using applied potential tomography Electron Lett ,19, 933-5
2. Kulkarni V, Hutchison J M S, Ritchie I K, Chesney R B , Gibson P , Mallard J R (1989) Monitoring fracture healing by impedance imaging EMBS 11th Annual International Conference
3. Kotre C J (1994) Electrical Impedance Tomography The British journal of Radiology, 70, S200-S205
4. Geddes L A, Baker L E (1967) The specific resistance of biological material - a compendium of data for the biomedical engineer and physiologist Med Biol Eng 5: 271-93
5. Kulkarni V, Hutchison J M S, Ritchie I K, Mallard J R (1989) Impedance imaging in upper arm fractures 29th Annual Scientific Meeting of the Biological Engineering Society, Bristol
6. Ritchie I K, Kulkarni V (1990) Impedance osteography: clinical applications of a new method of imaging fractures. J Biomed Eng 12:369-74
7. Hutchison J M S , Kulkarni V (1995) A novel 16-Electrode impedance imaging system Dept of Biomed Phys & Bioeng, Aberdeen Univ Conference: Innovations in Instrumentation for Electrical Tomography, IEEE Colloquium. DOI:10 1049/ic:19950644
8. Jongschaap H C N , Wytch R , Hutchison J M S , Kulkarni V (1994) Electrical impedance tomography: a review of current literature European journal of Radiology 18 165-174
9. Kulkarni V, Hutchison JMS, Mallard JR (1989) The Aberdeen impedance imaging system In: Carlson D, Ed Biomedical Sciences Instrumentation North Carolina: Instrumentation Society of America, 2 5: 47-58
10. The 6502 Second Processor User Guide. BBC Microcomputer (1984). Microprocessor System Editions.
11. Harikumar R , Prabu R , Raghavan S (2013) Electrical Impedance Tomography (EIT) and Its Medical Applications: A review Internacional Journal of Soft Computing and Engineering ISSN: 2231-2307, Volumen-3, Issue-4

Antonio H. Dell'Osa. Biomedic engineer graduated from Favaloro University in Buenos Aires, 2014. He has teached in mathematics area since 2009 and independent research since 2012. In 2015, he has incorporated in the research field in the National University of Tierra del Fuego, in Ushuaia.
Institute: Universidad Nacional de Tierra del Fuego.
Street: 879 H. Yrigoyen Street.
City: Ushuaia, Tierra del Fuego.
Country: Argentina.
Email: ahdellosa@untdf.edu.ar

Audio Codec and Digital Signal Processor for an Electrical Impedance Tomography System

N. Alfaro, M. Arregui, F. Martinucci, E. Santos, and F. Simini

Núcleo de Ingeniería Biomédica, Facultades de Medicina e Ingeniería, Universidad de la República, Uruguay

Abstract— **Electrical Impedance Tomography (EIT) can estimate thorax fluid/air content and distribution. Its use in critically ill patients is promising and may prove clinically useful. Currents are injected and voltages measured in the thorax boundary. The measurements are used to reconstruct tomographic images. An EIT system was implemented based on evaluation board OMAP-L137 (Spectrum Digital) and Howland current source. The board main components are the OMAP-L137 processor with DSP C6747 and AIC 3106 audio codec (Texas Instrument). Fewer than 5% difference in voltages against the oscilloscope was measured in the range between 42mV and 1.5V. The entire system showed 8.53% difference when measuring the real part of impedance.**

Keywords— **Electrical Impedance Tomography, Audio Codec, Digital Signal Processor, Synchronous Demodulation, Howland Current Source**

I. Introduction

Estimation of alveolar fluid content and distribution is essential in the management of conditions such as cardiogenic pulmonary oedema, pleural effusions, pneumonia and acute respiratory distress syndrome (ARDS). Electrical impedance of tissues can be estimated by measuring voltages on the skin while applying high frequency currents (> 10 kHz) whose amplitudes (< 5 mA) are below perception thresholds. Processing electrical impedance matrices yields tomographic images. This method, known as Electrical Impedance Tomography (EIT), is a low-cost, non-invasive, continuous-measurement method used to obtain low resolution images of the distribution of pleuro-pulmonary fluids and air. Figure 1 shows an EIT system typical architecture.

Since the pioneering work of Barber and Brown in 1984 [1], the last three decades have seen a considerable growth of EIT applications in research centers, with few commercial offers, notably Swisstom Pioneer system (Swisstom AG, Landquart, Switzerland)and PulmoVista 500 (Drägerwerk AG, Germany [2]). Since 1995 the Núcleo de Ingeniería Biomédica (nib) has developed circuits [3], reconstruction softwares [4] and complete prototypes [5, 6] under the name of IMPETOM (**impe**dance **tom**ography) with test results in phantoms and healthy volunteers.

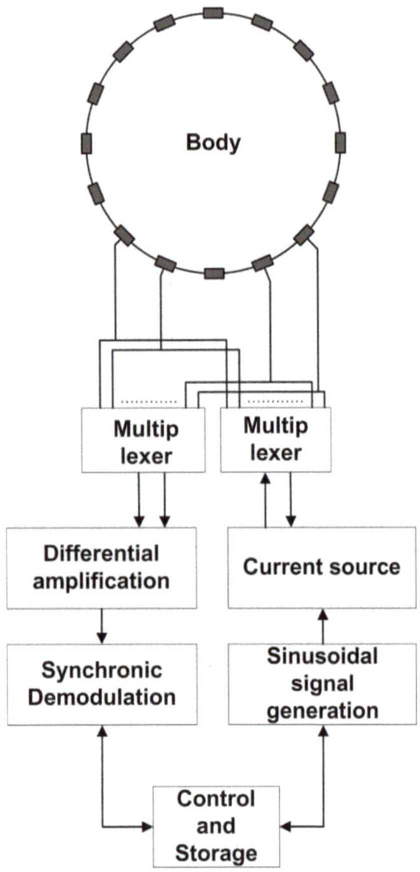

Fig. 1: EIT system basic block structure with a differential current source. There are 16 electrodes affixed onto the skin of the patient's body

The systems developed in recent years by research groups largely share some methods, such as direct digital synthesizer (DDS) to generate the sinusoidal signal or a configuration of simple or modified Howland as current source. In the measurement stage, the instrumentation amplifier is followed by an analog to digital converter of at least 16 bit resolution, then, in the digital domain the filtering, synchronous demodulation and other data processing is performed. For control and data processing the DSP is the most popular option for its speed and calculation capacity. In addition, the fewer mul-

F. Simini and P. Bertemes-Filho (eds.), *II Latin American Conference on Bioimpedance*,
IFMBE Proceedings 54,
DOI: 10.1007/978-981-287-928-8_5

tiplexers are used, the better the performance of the system. This setting seems to give the best compromise between performance and cost. An extended review on those methods can be found in [7]

This paper assesses the use of the Evaluation Module OMAP L137 from Spectrum Digital as core part of an EIT system. Particularly the audio codec AIC 3106 as signal generator and voltage measurements, and DSP C6747 from Texas Instruments for data processing.

II. SYSTEM DESCRIPTION

The EVM OMAP-L137 is an independent platform that allows developing and studying several applications for the OMAP-L137 integer circuit. It has incorporated a multi-core system on-chip (SoC) that permits to work simultaneously the two processors nested within: The floating point Digital Signal Processor (DSP) C6747 VLIW with 32bits, running at 300 Mhz and the RISC processor ARM926EJ-S also running at 300 Mhz (both from Texas Instrument). Besides, the EVM has a series of microcontrollers, peripherals, inputs and outputs for diverse purposes. We use the audio codec AIC3106 wich has two separate input channels with a resolution of 24 bits and 48 kHz sampling rate, the multichannel serial port McASP, the stereo inputs and outputs Line IN and Line OUT, and 32Megabytes of SDRAM. Figure 2 shows the EVM block diagram.

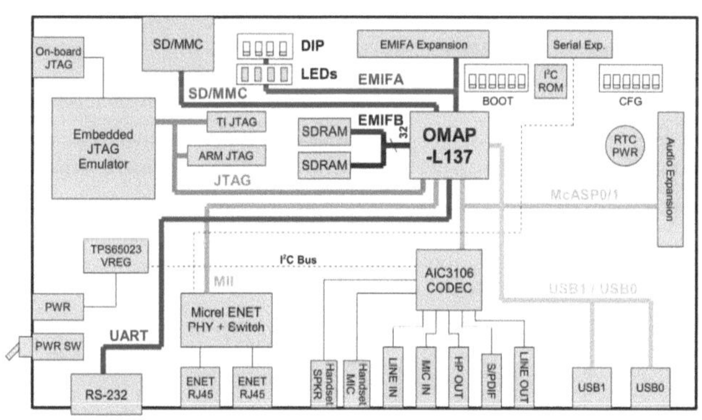

Fig. 2: Block Diagram OMAP-L137 EVM, taken from the module's Technical Reference Manual [8]

A. System Implementation

The system implemented is shown in figure 3. A 16 kHz digital sine wave is generated using a table within the DSP and sent to the Audio Codec through the serial channel McASP. The internal Digital to Analog Converter (DAC) in

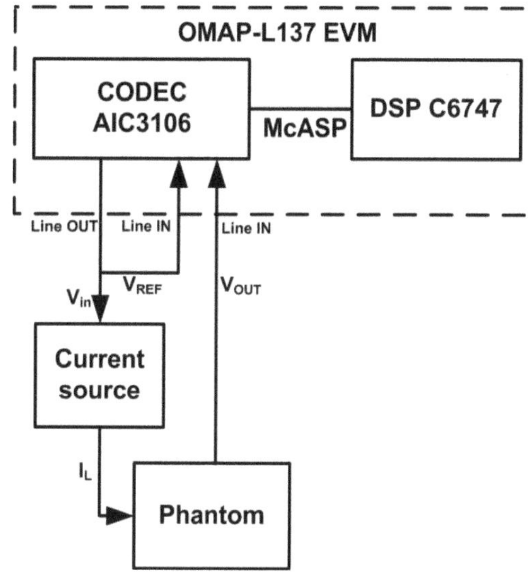

Fig. 3: Schematic block diagram for the OMAP L137 EVM based EIT System.

the AIC3106 transmits the converted signal through the Line OUT port. The voltage signal V_{in} feeds a Howland current source (figure 4). The source's output impedance depends on

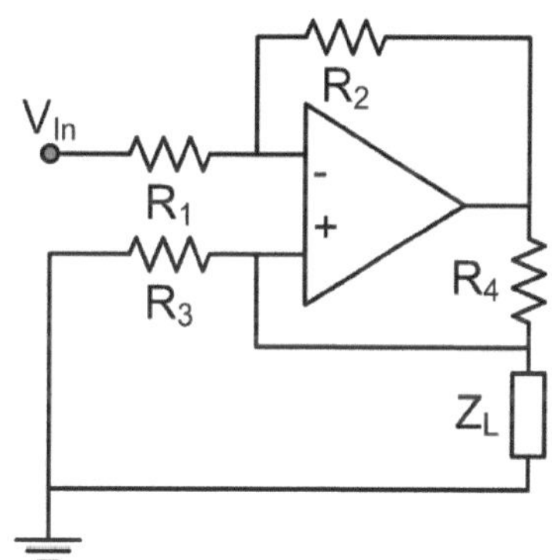

Fig. 4: Howland current source with operational OPA227.

the matching of the resistors and the operational's nonlinearities. In an ideal situation:

$$\frac{R_1}{R_2} = \frac{R_3}{R_4}, \tag{1}$$

and the output current, load independent, is given by the equation:

$$I_L = -\frac{V_{in}}{R_3}. \tag{2}$$

Input signals to the audio codec V_{REF} and V_{OUT} are amplified by an Programmable Gain Amplifier (PGA). Then signals pass through an antialiasing filter. Resultant signals are converted to digital samples at a sample rate of 48 kHz by means of an ADC that utilize sigma-delta technique. These samples are sent from AIC3106 to DSP C6747 through McASP channel.

A synchronous demodulator is implemented in the DSP in order to obtain the real part of the voltage V_{OUT} (figure 2). After digitizing, both input signals are multiplied sample to sample and inverted (because the current source has an inverter configuration). The real part of V_{OUT} is proportional to the DC component of the resulting signal. An Infinite Impulse Response (IIR) filter is used to filter the signal and the result corresponds to the DC component.

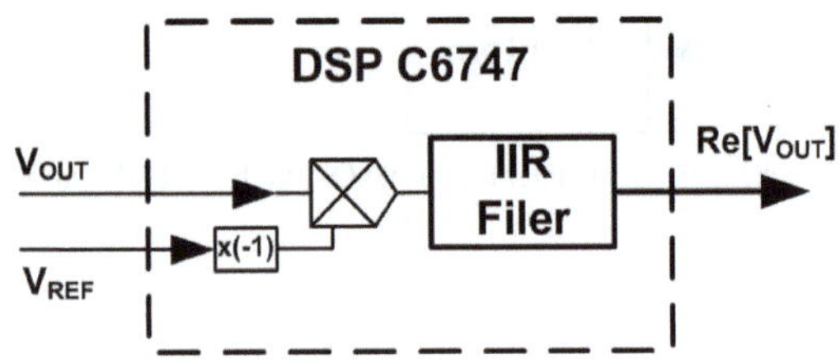

Fig. 5: synchronous demodulator implemented in the DSP to obtain V_{OUT} real component.

III. RESULTS

First we wanted to test the audio codec behavior as the ADC input channel. We used a Tektronix CFG253 signal generator directly connected to one of the Line in channels in the EVM. The sinusoidals input signals were simultaneously measured with oscilloscope Tektronix TDS210. The results show a linear response from the codec (figure 6), with differences under 5% between measured values in the range from 42 mV and 1.5 V.

Then the system in figure 3 was implemented with the Howland current source described before. A RC parallel circuit worked as our load impedance. An 100 nF capacitor was in parallel with a variable resistor. An algorithm to obtain the real part of the impedance was realized in the DSP given the linearity between Vref and the current generated. Figure 7 compares the DSP's estimated real part from the data measured with the codec with theoretical values and values obtained from the data measured with the oscilloscope. The EVM responds correctly to changes in the load impedance.

The mean difference between values estimated at the EVM with those based on the oscilloscope measurements was 8.53%. Every measurement was repeated 5 times to assess repeatability, with a result of 0.34% of mean deviation.

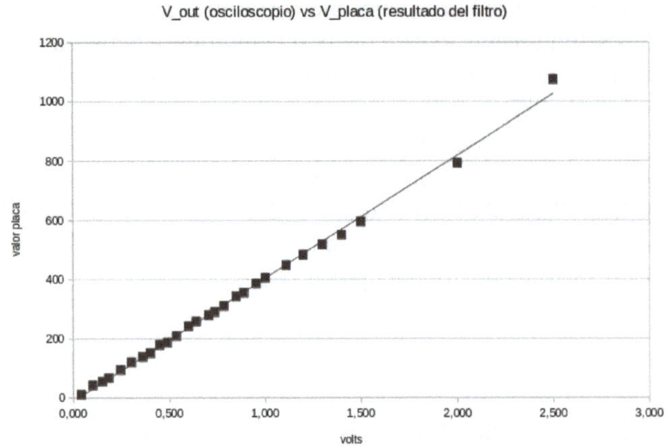

Fig. 6: Audio codec's Line In channel response to sinusoidal signals directly connected. Raw values displayed in the y-axis show the linearity between input and output. A simple gain adjustment is necessary to couple the signals.

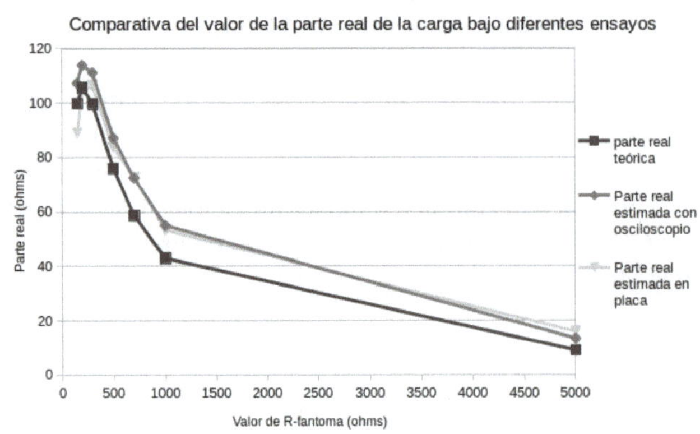

Fig. 7: Comparison of real part estimated in EVM, Oscilloscope and theoretical for a RC parallel circuit with 100 nF capacitor and variable resistor.

IV. CONCLUSION

The EVM OMAP-L137 with audio codec AIC3106 has proven apt to be an EIT system core. The system's measurement block shows a linear response for a broad range of voltages. The system as a whole, with a current source, showed bigger differences with respect to the estimation with the oscilloscope. Improving the current source is essential. Since EIT is a method of differences, the current should be such to provoke that the minimum voltage difference that we would like to distinguish (corresponding to the minimum changes in resistivity within the body) be greater than 42mV.

CONFLICT OF INTEREST

The authors declare that they have no conflict of interest.

REFERENCES

1. Smith R W, Freeston I L, Brown B H. A real-time electrical impedance tomography system for clinical use–design and preliminary results. *IEEE transactions on bio-medical engineering.* 1995;42:133–40.
2. Dräger . Dräger PulmoVista 500 Data Sheet 2011.
3. Ferreira A, Rodríguez A, Simini F. *IMPETOM-C Tomógrafo de Impedancia Eléctrica. Academic thesis* . 2002.
4. Hartman R, Lobo J, Ruétalo M, Simini F. *IMPETOM-I Tomógrafo de Impedancia Eléctrica. Academic thesis* . 2002.
5. González S, Liguori A, Simini F. *IMPETOM. Academic thesis* . 2005.
6. Quinteros Walter, Simini Franco. *IMPETOM-48 Tomógrafo de Impedancia Eléctrica con tres hileras de electrodos.* Academic thesisUniversidad de la República, Montevideo 2007.
7. Santos Eduardo, Simini Franco. Electrical Impedance Tomography for pulmonary oedema extent monitoring: review and updated design in *CLABIO - First Latin American Conference on Bioimpedance*(Joinville) 2012.
8. Spectrum Digital Incorporated. OMAP-L137 Evaluation Module Technical Reference 2008.

Author: E. Santos, F.Simini
Institute: nib-ncleo de ingeniera biomdica
Street: Hospital de Clnicas piso 15 sala 2, Av. Italia S/N, 11600
City: Montevideo
Country: Uruguay
Email: esantos@fing.edu.uy

Segmental Electrical Bioimpedance Measurements with a Single Lead (Electrode) Displacement

E. Colina-Gallo[1], C.A. González-Correa[2], C. Dussán-Lubert[3], and D.A. Miranda-Mercado[4]

[1] Universidad de Caldas, Department of Human Physical Activity, Manizales, Colombia
[2] Universidad de Caldas, Research Group on Electrical Bio-Impedance, Manizales, Colombia
[3] Universidad de Caldas/Department of Mathematics, Manizales, Colombia
[4] Universidad Industrial de Santander/Department of Physics, CIMBIOS, Bucaramanga, Colombia

Abstract— **Segmental Bioelectrical Impedance Analysis (SBIA) has been proposed as an alternative to overcome some limitations of total BIA measurements for the calculations of body composition. It also has been used in different clinical conditions such as ascites, edema, neuromuscular damage, and during hemodialysis. There is no agreement about the best electrode arrangement to carry out the measurements. Using a sample of 43 apparently healthy adults recruited at the university (16 men aged 17-28 y, and 27 women aged 17–50 y), we show how, with the displacement of just one single voltage electrode, we can get readings that give accurate results and are less time consuming. For this purpose, we compared the results obtained through our proposal with those obtained when using three different methods reported in the scientific literature. We also propose a standardized nomenclature for electrode positioning.**

Keywords— **Segmental bioimpedance, appendicular resistances, electrode array.**

I. Introduction

Bioelectrical impedance (BI) measures the opposition (impedance) of biological tissues to the flow of alternating electric current. This opposition is highly correlated both with the concentrations of electrolyte ions and water in the tissues, as well as with their structure: cell sizes, density, and spacing [1] . In the field of body composition, the technique is known as Bioelectrical Impedance Analysis (BIA), and it has been used for the estimation of total body water (TBW), fat-free mass (FFM) and fat mass (FM) [2]. Impedance is measured using a tetrapolar approach as described in [3].

Since the late 1980s, Segmental Bioelectrical Impedance Analysis (SBIA) has been used to overcome some limitations of the whole body BIA approach, arising from the small contribution of the trunk to whole body resistance and the assumption of representing the human body as a conducting cylinder with uniform cross-sectional area [4] SBIA measures the resistance of upper limb, lower limb and trunk in order to calculate total FFM. It also allows the analysis of body composition of specific body segments in patients with conditions like ascites, local or regional edema, and tissue deposition or depletion [5]. SBIA has also been used for estimation of hydration during hemodialysis, monitoring postsurgical edema, lymphedema and neuromuscular disease [5].

Different arrays have been proposed for SBIA, especially in relation to the positioning of current and voltage electrodes. In their approach, Bracco *et al* [6], kept the current injecting electrodes at the same place as for whole body analysis. Some investigators like Chumlea *et al* [7], moved the current electrodes and placed them near the detecting electrodes. Organ *et al* [4], proposed that "virtual" limb resistance can be measured using the contralateral wrist and ankle as the electrode sites for the upper limb/trunk and trunk/lower limb, respectively. In their attempt to obtain a resistance profile along the limbs, Stahn *et al* [8], used current electrodes in standard position and voltage electrodes at 2 cm intervals to the proximal part of the femoral head. Resistance was determined between the original electrodes at the wrist and each electrode along the lower limb, and the measured resistance was subtracted from the whole body resistance to obtain the resistance profile of the lower limb. A similar approach was used to determine a resistance profile of the upper limb.

When SBIA is used, it would be expected that, if the whole body is divided into segments, the measurements should comply with the additivity principle, that is:

$$R_{Total} = R_1 + R_2 + \dots + R_n, \quad (1)$$

in our case,

$$R_{Total} = R_{Arm} + R_{Trunk} + R_{Leg.} \quad (2).$$

The additivity principle applies because as the current flows between both extremes of the body, when segmental voltages are measured, it is assumed that it has a uniform distribution across each segment volume and the potential differences follow the Kirchhoff's voltage law. Although most of the different approaches used for SBIA give very similar results, some of them do not exactly comply with this principle. In this report, it is shown that just by displacing one single lead (i.e., moving it to the electrode placed in a different location), in a similar way as Stahn *et al* [8], the method complies with the principle and is also less time consuming when compared to the need of moving two or more leads. One can also be confident that the obtained values do reflect the resistance of the measured segments. In this study, this approach is compared with the results obtained by three other methods already mentioned: a) Chumlea *et al* [7], b) Organ *et al* [4], c) Bracco *et al* [6].

F. Simini and P. Bertemes-Filho (eds.), *II Latin American Conference on Bioimpedance*,
IFMBE Proceedings 54,
DOI: 10.1007/978-981-287-928-8_6

Finally, we also propose a standardized nomenclature for electrode positioning when using SBIA.

II. MATERIALS AND METHODS

A. Subjects

The study was carried out with 43 apparently healthy adults, 16 men (aged 17-28 y) and 27 women (aged 17–50 y), recruited at the University of Caldas. Weight and body fat percentage were measured with a TANITA analyzer (BC-418 model). Stature was measured with a meter wall. Their physical characteristics are summarized in Table 1.

Table 1 Physical characteristic of subjects $\bar{x}(SD)$

	Age (years)	Weight (kg)	Height (m)	BMI (kg/m^2)	% BF
Men (n=16)	20,4 (3,8)	67,3 (10,3)	1,70 (0,04)	22,2 (2,7)	16,3 (3,8)
Women (n=27)	23,9 (10,9)	58,5 (8,9)	1,59 (0,06)	23,2 (3,1)	29,0 (5,1)

B. *Bioimpedance measurements*

Whole body and segmental measurements of raw resistance (R) were performed at 50 kHz, with a nominal current of 400μA RMS, and using a 4000B Bio-impedance spectrum analyzer system by XITRON technologies (San Diego, USA). Subjects were asked to void their bladder and remain supine five minutes before the measurements. Measurements were carried out by the same operator (ECG) and they were completed within ten minutes. These were performed with the subject lying supine on a non-conducting surface and with arms and lower limbs abducted 45°. Bioelectrical impedance was measured on the right side of the body, from hand to shoulder (upper limb), from shoulder to groin (trunk), and from groin to foot (lower limb). Total body impedance was measured from hand to foot. EKG electrodes (2228 by 3M ®) were used [9], and the sites where they were going to be placed were cleaned with alcohol swabs before their application. For the identification of the electrode sites, we proposed the nomenclature schematized in Fig. 1. Basically, we distinguished between right (R) and left (L) sides, as well as extra (E) positions. The sites for the conventional current electrodes were named with the corresponding letter for the side (either R or L), and a two digit number: 00 for hand (distal metacarpus) and 99 for the foot (distal metatarsus). We have, therefore, the names R00, R99, L00 and L99, respectively. It can also be seen that, for sensing electrodes, odd numbers correspond to joints, while even numbers correspond to areas between them. Although, in this article, we only considered the three large segments so far mentioned (upper limb, trunk and lower limb) we think that, in the future, this nomenclature will be useful, especially when smaller segments are measured.

For the measurement of whole body impedance, electrodes were placed as described in [10]. Four different arrangements were used to measure the bioelectric impedance of upper limb, trunk and lower limb. In the first arrangement, electrodes were arranged according to the description given by Chumlea *et al* [7]. For the second array, we followed the protocol for virtual segmental measurements described by Organ *et al* [4]. The third arrangement was the one described by Bracco *et al* [6]. In the fourth arrangement, our proposal, current electrodes and the hand (voltage) electrode remained in the same position as for whole body impedance, while the lead of the foot (voltage) electrode was successively moved from the ankle to the thigh and from the thigh to the shoulder. Impedance values of the segments were obtained by subtraction from total impedance. The only difference between our proposed method (method 4) and the Bracco *et al* [6] method (method 3 in this paper) is that, instead of reconnecting the two voltage leads only one voltage lead must be changed in the new method. Electrode positions for each arrangement are summarized in Table 2.

C. Data analysis

A statistical analysis was performed using XLSTAT 2014. The correlation between the readings made by the three and four methods were analyzed using the intraclass correlation coefficient (ICC) [11] and the graphical method of Bland and Altman [12].

Table 2 Electrode positions for segmental measurements for four different arrangements

Arrangement	Upper limb current	Upper limb voltage	Trunk current	Trunk voltage	Lower limb current	Lower limb voltage
1. Chumlea *et al* (1988),	R00 – E03	R01 – R05	E01 – E05	E02 – R07	R99 – E04	R07 – R11
2. Organ *et al* (1994)	R00 – R99	R01 – L01	R00– R99	L01 – L11	R00 – R99	R11 – L11
3. Bracco *et al* (1996),	R00 – R99	R01 – R05	R00– R99	R05 - R07	R00 – R99	R07 – R11
4. Our proposal	R00 – R99	R01 – R05	R00– R99	R01- R07 minus R01- R05	R00 – R99	R01- R11 minus R01-R07

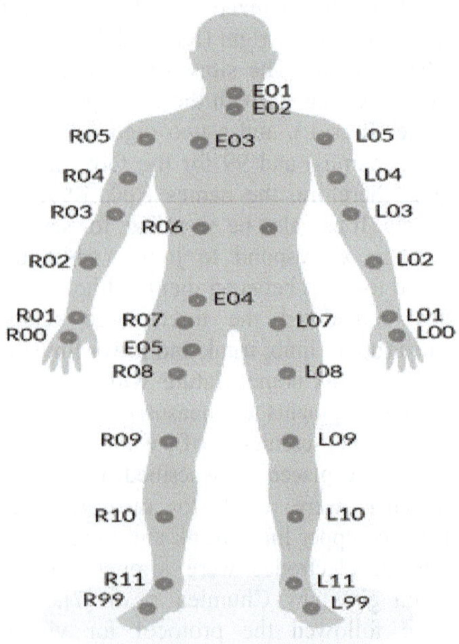

ELECTRODE	POSITION
R00/L00	Posterior surface of hand at distal end of third metacarpal.
R01/L01	Posterior surface of wrist, between radius and ulna.
R02/L02	Midpoint between radial process and olecranon.
R03/L03	Just over lateral epicondyle.
R04/L04	Midpoint between olecranon and acromial process.
R05/L05	Acromial process.
R06/L06	Lower costal margin on medial axillar line.
R07/L07	Anterior surface of thigh on the midline, at gluteal crease.
R08/L08	Midpoint between R07/L07 and upper edge of patella.
R09/L09	Just over external joint line of knee.
R10/L10	Midpoint between R09/L09 and external malleoli.
R11/L11	Anterior surface of the ankle, between internal and external malleoli.
R99/L99	Anterior surface of the ankle, between internal and external malleoli.
E01	5,0 cm proximal to sternal notch, in the midline on anterior surface of the neck.
E02	At sternal notch.
E03	5,0 cm medial to R05.
E04	5,0 cm proximal to R07
E05	5,0 cm distal to R07

Fig. 1 Proposed standardized nomenclature for the anatomical positions of electrode placements

III. RESULTS

There is no statistically significant difference between mean values obtained by the four methods in upper limb, lower limb and the sum of resistances, both in men and women. A summary statistics of the segmental sum obtained by the four methods are given in Table 3. Trunk resistances show significant differences among all methods except between that of Bracco *et al* [6] and our proposal, both in men and women.

Table 3 Segmental sum (Ω) by four methods $\bar{x}(SD)$

	Chumlea *et al* (1988)	Organ *et al* (1994)	Bracco *et al* (1996)	Our proposal
Men (n=16)	554,2 (58,2)	534,4 (61,8)	531,4 (59,7)	528,0 (59,1)
Women (n=27)	667,8 (62,5)	634,3 (68,4)	635,4 (63,1)	629,3 (62,7)
Total (n=43)	625,5 (82,0)	597,1 (81,5)	596,7 (79,5)	591,6 (78,3)

The intraclass correlation coefficient (ICC) estimates the average correlation between all possible pairs of observations [13]. The ICC between methods 3 and 4 indicates, according to Fleiss JL [11], a very good agreement between them in arm, leg, trunk and in the segmental sum. When using the graphical method of Bland and Altman [12] mean values of upper limb resistances obtained by method 3 and 4 are not different, trunk and lower limb resistances, and the sum of the segments made by method 3 always show superior readings than method 4 (differences in all cases were positive); the discrepancies are stable or homogeneous when higher values of mean differences are obtained, as showed in the example given in Fig. 2. The comparison between the differences and the means of the obtained values show no correlation (r^2= 0,18).

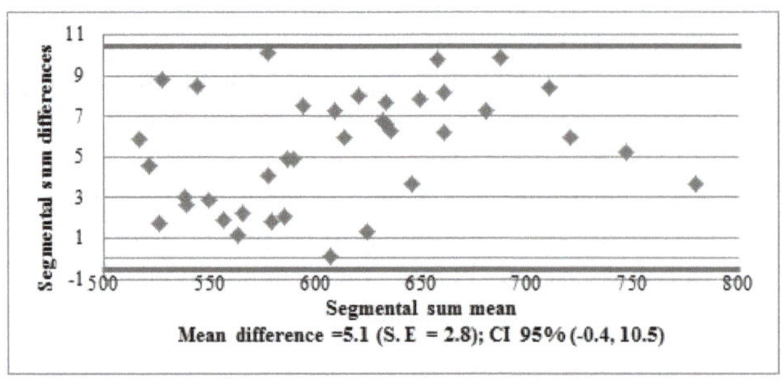

Fig. 2 Bland and Altman plot. Segmental sum mean resistance (Ω) between methods 3 and 4

IV. DICUSSION

The Organ *et al* [4] method overestimates upper limb

resistances about 8% (probably because arm resistance includes a contribution from the upper trunk), consistent with those reported by other authors[14]. Trunk resistances show significant differences among all methods, a fact already reported in the literature, as the trunk is the major source of discrepancy in segmental measurements [15]. Mean values of the sum of resistances are no different by the four methods. The mean of the sum of the three segments is 5.5% higher than the total with the Chumlea *et al* [7] method. This average is lower than the one calculated from the author's data (14% in men and 13% in women) [7]. With the Organ *et al* [4] and the Bracco *et al* [6] methods, the difference is about 1%. Our proposal shows a 100% concordance with the additive principle, i.e., that the sum of the segments must be equal to the total resistance. The values of trunk resistances are the same irrespective of the direction of electrode displacement, i.e., R01-R07 minus R01-R05 is equal to R11-R05 minus R11-R07. This was tested in 4 subjects (data not reported).

The intraclass correlation and graphical analysis of Bland and Altman show good agreement between the method described by Bracco *et al* [6] and our proposal.

V. Conclusion

In summary, despite the small sample size and that only persons with normal body composition were used for our experiments, our proposal suggests that, with the displacement of a single lead to different voltage electrodes it is possible to measure not only the electrical impedance of the three major segments traditionally used in SBIA, but also of smaller segments, when required. Our method is similar to the method proposed by Bracco *et al* [11], but is less time and electrode consuming. It is necessary to make further comparisons between our method and the method proposed by Bracco *et al* [11] in subjects with different conditions of ascitis or edema, as with as multifrequency analysis. We hope that the standardized nomenclature proposed in this article can be of help for other authors.

Compliance with Ethical Requirements

All subjects gave their informed written consent in accordance with the guidelines of the Ethical Committee of the University of Caldas (Manizales, Colombia).

Conflict of Interest

The authors declare that they have no conflict of interest.

References

1. Schwan H, Kay C (1956) Specific resistance of body tissues. Circ Res 4(6):664–70 DOI 10.1161/01.RES.4.6.664
2. Fuller N, Elia M (1989) Potential use of bioelectrical impedance of the "whole body" and of body segments for the assessment of body composition: comparison with densitometry and anthropometry. Eur J Clin Nutr 43:779–91
3. Lukaski H, Bolonchuk W, Hall C et al (1986) Validation of tetrapolar bioelectrical impedance method to assess human body composition. J Appl Physiol 60(4):1327–32
4. Organ L, Bradham G, Gore D et al (1994) Segmental bioelectrical impedance analysis : theory and application of a new technique. J Appl Physiol 77(1):98–112
5. Ward L (2012) Segmental bioelectrical impedance analysis: an update. Curr Opin Clin Nutr Metab Care 15(5):424–9 DOI 10.1097/MCO.0b013e328356b944
6. Bracco D, Thiébaud D, Chioléro R et al (1996) Segmental body composition assessed by bioelectrical impedance analysis and DEXA in humans. J Appl Physiol 81:2580–7
7. Chumlea W, Baumgartner R, Roche A (1988) Specific resistivity used to estimate fat-free mass from segmental body measures of bioelectric impedance. Am J Clin Nutr 48:7–15
8. Stahn A, Terblanche E, Strobel G (2007) Modeling upper and lower limb muscle volume by bioelectrical impedance analysis. J Appl Physiol 103(4):1428–35 DOI 10.1152/japplphysiol.01163.2006
9. Caicedo-Eraso J, González-Correa C, González-Correa C (2012) Use of electrocardiogram (ECG) electrodes for bioelectrical impedance analysis (BIA). J Phys Conf Ser 407:012008 DOI 10.1088/1742-6596/407/1/012008
10. Lukaski H, Johnson E, Bolonchuk W et al (1985) Assessment impedance of fat-free mass using bioelectrical measurements of the human body. Am J Clin Nutr 41:810–7
11. Fleiss J (1986) The design and analysis of clinical experiments. Wiley, New York DOI 10.1016/0898-1221(87)90245-8
12. Bland J, Altman D (1986) Statistical methods for assessing agreement between two methods of clinical measurement. Lancet 1:307–10
13. Bland J, Altman D (1996) Measurement error and correlation coefficients. BMJ 313(7048):41–2 DOI 10.1136/bmj.313.7049.106
14. Jaffrin M, Bousbiat S, Dongmo E (2011) A comparison between two methods for measuring limb resistances with wrist and ankle electrodes. Med Eng Phys 33(8):943–9 DOI 10.1016/j.medengphy.2011.03.003
15. Baumgartner R, Chumlea W, Roche A (1989) Estimation of body composition from bioelectric impedance of body segments. Am J Clin Nutr 50:221–6

Author: Evelyn Colina Gallo
Institute: Universidad de Caldas
Street: Calle 25 #48-57
City: Manizales
Country: Colombia
Email: evelyn.colina@ucaldas.edu.co

An Alternative Electrical Impedance Myography Technique for Assessment of Local Muscular Fatigue

A.B.B. Coutinho, B. Jotta, T.S. Carvalho, A.V. Pino, and M.N. Souza

Universidade Federal do Rio de Janeiro/COPPE, Biomedical Engineering Program, Rio de Janeiro, Brazil

Abstract—**Electrical impedance myography (EIM) is a technique that has been used to evaluate muscle contraction features, but its application to assess fatigue is not completely described yet. This work presents the comparison between the detection of muscle fatigue by the muscle force and its counterpart performed by bioelectrical impedance. The experimental protocol performed by five volunteers was consisted by isometric contractions of the elbow flexor complex (biceps brachii, brachialis, and brachioradialis muscles) at 30% of their maximum voluntary contraction (MVC) up to the exhaustion. Time series of muscle force and bioelectric impedance data were obtained for a sample period of 10 seconds. Bioelectric impedance data were acquired by using a bipolar bioelectric impedance spectroscopy method based on the current response to a voltage step excitation (BIS-STEP). The findings indicate that the used impedance measurement method was sensible enough to detect changes on the muscular bioelectrical impedance that were associated to fatigue, suggesting an early detection of fatigue by the bioelectric impedance data when compared with detection performed by force data.**

Keywords— **Electrical Impedance Myography, Muscle Contraction, Muscle Fatigue**

I. Introduction

There are different techniques used to estimate fatigue of a muscle during activity, where probably one of the oldest and well known is one based on Electrical Myography (EMG) [1]. More recently, a technique called Electrical Impedance Myography (EIM) has been applied to investigate features of muscle contraction through changes at its electric characteristics [2]. These electrical changes are related to the biochemical processes that occur during muscle contraction. Considering an isometric contraction of a healthy muscle, from the beginning until fatigue, those biochemical processes may interfere on the characteristics of the muscle electric impeditivity.

Some works at literature have reported small changes in muscle conductivity during contraction [2], suggesting low sensitivity of bioelectric impedance measurement techniques to assess the impeditivity changes caused by muscle activity. Additionally, few studies tried to investigate bioelectric impedance parameters associated to muscle fatigue. Isometric contractions might be a good strategy to investigate fatigue features because such contractions increase the intramuscular pressure, favoring to a local accumulation of metabolites [3].

The technique used to measure bioelectric impedance in some previous EIM studies might have contributed to the low sensitivity reported [1], where usually a tetrapolar arrangement of electrodes was involved in order to avoid influences of electrode impedance. However, in a previous work [4], a bipolar bioelectric impedance spectroscopy method based on the current response to a voltage step excitation (BIS-STEP) [5] was used in an EIM investigation leading to strong correlations between impedance parameters and muscle strength.

The present paper studied the use of BIS-STEP technique during a fatigue protocol of biceps brachii muscle through isometric contraction, aiming to know if the mentioned measurement technique exhibits enough sensitivity to detect impedance changes provoked by metabolic events.

II. Methods

Instrumentation

A calibrated load cell (SV200, Alfa Instrumentos, Brazil) fixed on a weight machine was used to measure the isometric contraction force of the elbow flexor complex (biceps brachii, brachialis, and brachioradialis muscles). The electric signal supplied by the analog circuit of the load cell was sampled at 200 Hz.

Bioelectric impedance data were acquired using a prototype device (hardware and software) that implements the BIS-STEP method [9]. Such a prototype used an AD converter with a sample rate of 1.25 Ms/s.

Two brass electrodes (12 cm² area) with electrolytic gel were used to minimize the effect of the impedance of the electrode-tissue interfaces during the bioelectric impedance data acquisition. The electrodes were placed 2 cm above and bellow the biceps brachii muscle belly. Removal of the *stratum corneum* layer by rubbing with cotton and saline solution was accomplished before electrodes placement.

B. Experimental Protocol

Initially a maximum voluntary contraction (MVC) was performed by the volunteer through isometric contraction of

F. Simini and P. Bertemes-Filho (eds.), *II Latin American Conference on Bioimpedance*,
IFMBE Proceedings 54,
DOI: 10.1007/978-981-287-928-8_7

elbow joint at a 90° angle. Three MVC were performed, with three minutes rest period between them, and the maximum value obtained was recorded. After an additional three minutes of resting period the volunteer began the test, which consisted of sustain an isometric contraction of elbow flexion at thirty percent of MVC (*MVC30%*) and at the same 90° angle. Bioelectric impedance data were acquired every 10s during muscle contraction. The test lasted until the intensity of the isometric contraction decreases ten percent (or less) of the force level corresponding to *MVC30%*.

C. Sample

Five male subjects, aging 25.4 (mean) ± 4.1 (standard deviation) years, height 1.74 ± 0.1 m, and weight 77.2 ± 18.3 kg participated in the study. All subjects were right-handed without previous registration of neuromuscular injuries at the upper limbs. The study was submitted and approved by the Ethic Committee of the University Hospital Clementino Fraga Filho (HUCFF) at Federal University of Rio de Janeiro, under the number 312.381.

D. Analysis and Data Processing

Bioelectrical Impedance Data: In order to reduce the random fluctuation in data acquisition, three experimental current responses associated to the BIS-STEP method was obtained through the average of 50 voltage steps. Figure 1 depicts a typical current response.

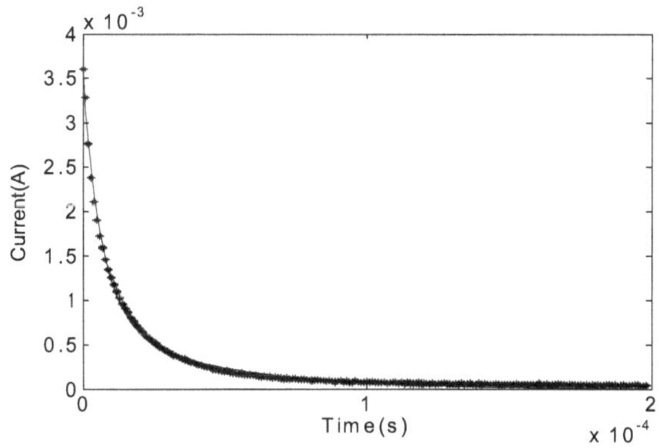

Fig 1. Example of experimental current response data obtained by the BIS-STEP method.

The electric model depicted in figure 2 was used to interpret the current response associate to BIS-STEP method. Due to the usage of electrodes with large surfaces, we considered the resistance Rb (associated to the electrode-tissue interface) negligible compared to the magnitude of bioelectric impedance. Thus, one can see that when all the capacitors are discharged (beginning of the step excitation) the current is limited by the parallel association of the intra-cellular resistance - Ri and extra-cellular resistance - Re, hereafter called Rinf. Such a parameter (estimated by equation 1) was used as the feature that summarizes the influences of intra and extra-cellular mediums on the current response.

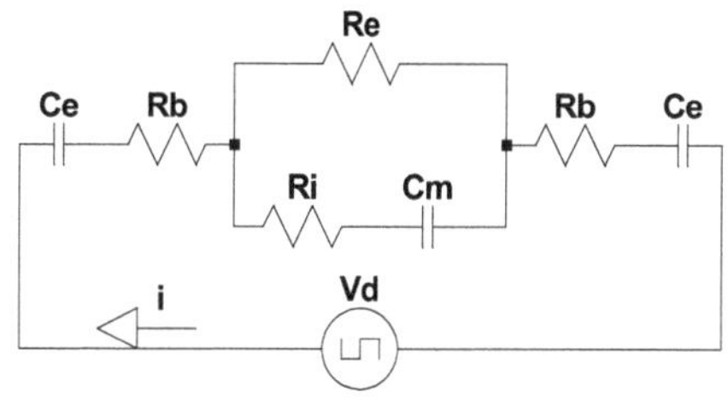

Fig. 2. Electric model used by the method BIS-STEP to interpret current response *i* to a step voltage of magnitude *Vd*. The extracellular resistance – *Re*, intracellular resistance – *Ri*, and membrane capacitance – *Cm*, stand for the biological segmental bioelectric impedance model. The capacitances – *Ce* and resistances – *Rb* are associated to the impedances of the two electrode-tissue interfaces.

$$R_{\inf} = \frac{Vd}{i_p} \qquad (1)$$

where Vd is the magnitude of step voltage and i_p is the highest value of current response.

The time series of $R_{\inf}$ was analyzed for the five subjects, in order to seek the influence of muscle fatigue on bioelectric impedance. To better investigate the time series of $R_{\inf}$ of the volunteers, we computed the percentile change of $R_{\inf}$ with respect to $R_{\inf}$ value at the beginning of muscular effort ($R_{\inf 0}$).

$$R_{\inf i\%} = \frac{(R_{\inf i} - R_{\inf 0})}{R_{\inf 0}} \times 100\% \qquad (2)$$

where, $R_{\inf i}$ is the $R_{\inf}$ value at time i, $R_{\inf 0}$ is the value of $R_{\inf}$ at $t = 0$.

Then, a moving difference digital filter (equation 3) was applied to the time series of the relative changes of $R_{\inf\%}$.

$$\Delta R_{\inf i\%} = R_{\inf i\%} - R_{\inf (i-1)\%} \qquad (3)$$

Force Data: As mentioned before, bioelectric impedance data were acquired every 10 s, i.e. equivalent to a sample frequency of 0.1Hz, and the force data were acquired at 200Hz.To obtain a time series of muscle force comparable to the time series of $\Delta R_{\inf\%}$, the force data sampled at 200Hz was low-pass filtered (4th order Butterworth, in forward and reverse order) at 0.05Hz, (equivalent to the Nyquist frequency for a sample frequency of 0.1Hz) and then resampled at 0.1 Hz. Then, the time series of muscle force was transformed to the time series of relative changes of force (equation 4).

$$F\% = \frac{(F_i - F_{initial})}{F_{initial}} \times 100 \quad (4)$$

The behavior of $\Delta R_{\inf\%}$ time series was observed and compared with the time series of *F%*. The fatigue threshold was defined as the time ($t_{(F_{end})}$) where the times series of relative muscle force (*F%*) exhibits a negative change with a magnitude greater than 10%. By observing the time series of $\Delta R_{\inf\%}$ we hypothesized that the fatigue threshold is associated to the time ($t_{(R_{\inf F})}$) where the series presents the first negative change presenting a magnitude greater than 0.6% (experimentally set) after one minute of contraction. The two estimates of fatigue threshold ($t_{(F_{end})}$ and $t_{(R_{\inf F})}$) were compared for all the volunteers.

All data processing was performed through MATLAB 7.2 environment.

III. RESULTS

Figures 3, 4 and 5 show the time series of $R_{inf\%}$, $\Delta R_{inf\%}$ and *F%* for the second volunteer, which presents the average general behavior of all the volunteers.

In figure 4 the output of moving difference filter (ΔR_{inf}) is plotted and the estimate of fatigue threshold $t_{(R_{\inf F})}$ is indicated by a dashed vertical line. Figure 5 shows the time series of the relative muscle force (*F%*) and the estimate of fatigue threshold $t_{(F_{end})}$ indicated by a dashed vertical line.

Table 1 shows the estimated of fatigue threshold obtained from times series of bioelectric impedance ($t_{(R_{\inf F})}$) and force ($t_{(F_{end})}$), the test period ($t_{(total)}$), and the time lag (Δt) between $t_{(F_{end})}$ and $t_{(R_{\inf F})}$, for all the five subjects.

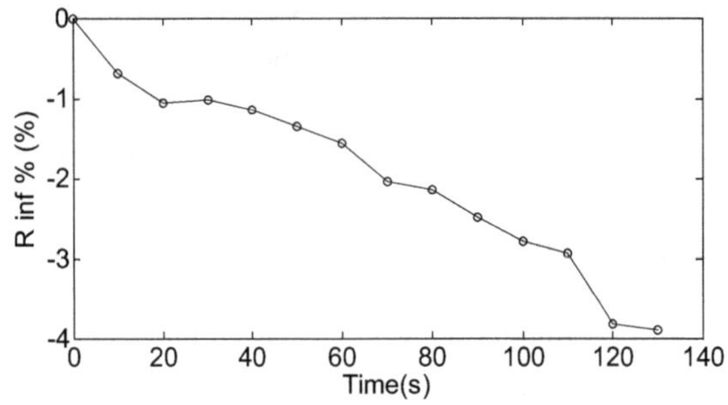

Fig. 3 Time series of relative changes of R_{inf} along the time of isometric contraction of the biceps brachii muscle.

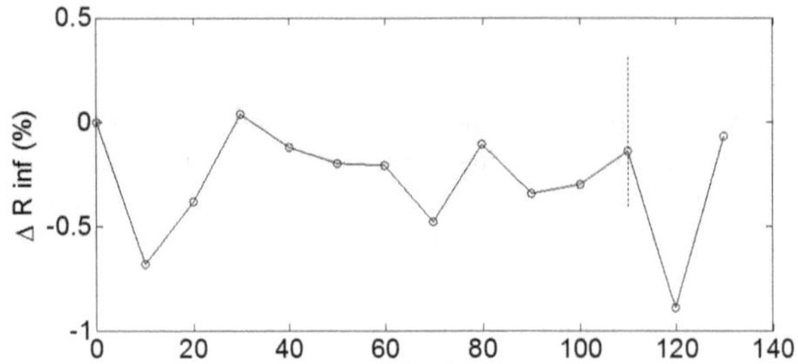

Fig. 4 Time series of $\Delta R_{in\%}$ along the time of isometric contraction of the biceps brachii muscle. The estimate of fatigue threshold $t_{(R_{\inf F})}$ is indicated by a dashed vertical line

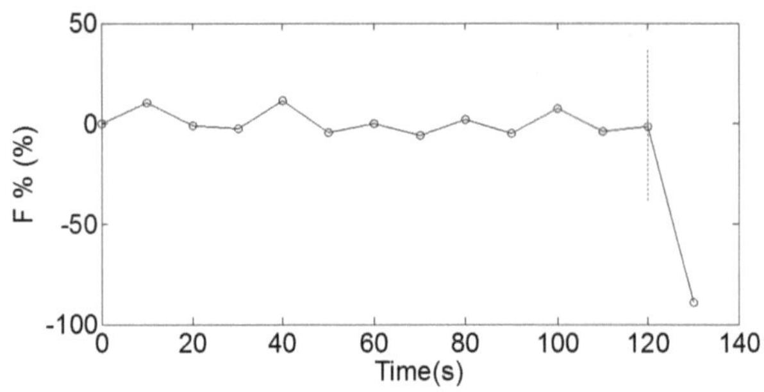

Fig. 5 Time series of relative force along the time of isometric contraction of the biceps brachii muscle. The estimate of fatigue threshold $t_{(F_{end})}$ is indicated by a dashed vertical line.

The largest time lag between $t_{(R_{\inf F})}$ and $t_{(F_{end})}$ was 30 seconds, and the biggest relative difference between the two estimates of fatigue thresholds was about 22% (volunteer 4). For the volunteers 1 and 2 such relative differences were about 8%.

Table 1 : Estimated of fatigue thresholds obtained by the bioelectric impedance ($t_{(R_{\inf F})}$) and by the force ($t_{(F_{end})}$) times series for each volunteer. The total length of the time series ($t_{(total)}$) is also shown, as well as the time lag (Δt) between $t_{(F_{end})}$ and $t_{(R_{\inf F})}$.

Volunteer	1	2	3	4	5
$t_{(R_{\inf F})}$ (s)	110	100	140	100	60
$t_{(F_{end})}$ (s)	120	110	170	120	70
$t_{(total)}$	130	120	180	130	80
Δt (s)	10	10	30	20	10

IV. Discussion

A general behavior of the time series associated to R_{inf}, $\Delta R_{\inf\%}$, and *F%* was observed for all the five volunteers.

The results showed some agreement between the two estimates of muscular fatigue and that $t_{(R_{\inf F})}$ was in general 10 to 30 s smaller than $t_{(F_{end})}$. It seems reasonable to think that biochemical changes associated to muscle fatigue could be firstly detected by bioelectric impedance [6], when one compares its mechanical consequence in muscular strength, which appears sometime after and that would be detected later by the force data.

The results are in agreement with some previous works [7], where the impedance decreases along the time of muscular contraction. However, these studies have used protocols considering dynamic contraction, where the blood flow is higher than in isometric contractions [6].

The results could have been influenced by different volunteers' abilities to tolerate the muscle pain provoked by fatigue process [8]. The present study did not considered this issue.

Despite using electrodes with 12 cm² of area and electrolytic gel to minimize the influence of the impedances associated to electrode-tissue interfaces, such impedances might have exerted some effect on the current response associated to the BIS-STEP method. Thus, it is possible that the observed changes of R_{inf} can be even higher than the ones registered in this first approach.

V. Conclusion

Even using just R_{inf} as a summarized parameter and some limitation of the experimental protocol, the findings indicated that the BIS-STEP method seems to be sensible to detect metabolic changes during muscle contraction. Further experiments must be conducted to assure the use of this technique in the assessment of muscle fatigue.

Conflict of Interest

The authors declare that they have no conflict of interest.

References

1. González-Izal M, Malanda, A, Gorostiaga E, et al (2012) Electromyographic models to assess muscle fatigue. J Electromyography Kinesiol, 22(4):501-512.
2. Shiffman, CA, Aaron, R, et al. (2003) Electrical impedance of muscle during isometric contraction. Physiol Meas 24(1):213.
3. Henriksson J, Katz A, Sahlin K. (1986) Redox state changes in human skeletal muscle after isometric contraction. J Physiol 380(1):441-451
4. Coutinho ABB, Jotta B, Pino AV, et al. (2012) Behaviour of the electrical impedance myography in isometric contraction of biceps brachii at different elbow joint angles. J Phys Conf Ser 407(1):012-017
5. Neves CEB, Souza MN (2000).A method for bio-electrical impedance analysis based on a step-voltage response. Physiol Meas 21(3):395
6. Silva OL, Sousa TH, Hoffman IO, et al (2014) A proposal to monitor muscle contraction through the change of electrical impedance inside a muscle, Biomedical Robotics and Biomechatronics5th IEEE RAS & EMBS International Conference, São Paulo, Brazil, 2014, pp 763-767
7. Ikai RL, Ryosukc S, Kyoso M et al. (2007) Dynamic Electrical Characteristics of Muscle with the Fatigue during Exercise, IFMBE Proceedings of World Congress on Medical Physics and Biomedical Engineering, South Korea, Seoul, 2006 pp 622-625
8. Jones DA, Newham DJ, Torgan C (1989) Mechanical influences on long-lasting human fatigue and delayed-onset pain. J Physiol 412(1):415-427.

Author: André Brand Bezerra Coutinho
Institute: Federal University of Rio de Janeiro
Street: 3020 Horácio de Macedo Avenue
City: Rio de Janeiro
Country: Brasil
Email:andrebrand1@hotmail.com

In Vitro Luminal Measurements of Colon Electrical Impedance in Rabbits

Edelberto Mulett-Vásquez[1,2], Amilbia Correa-Florez[1], Cármen Dussán-Lubert[3], David-Alejandro Miranda-Mercado[4], and Carlos-Augusto González-Correa[1]

[1] Research Group on Electrical Bio-Impedance. University of Caldas, Manizales, Colombia, South America
[2] Department of Surgery. University of Caldas, Manizales, Colombia, South America
[3] Department of Mathematics, University of Caldas, Manizales, Colombia, South America
[4] CIMBIOS, Santander Industrial University, Bucaramanga, Colombia, South America

Abstract. **Until very recently, the colon was a neglected organ in science. During the present century, it has gained much attention, due to its role in health and disease, especially because of the presence of the microbiota. The interaction between the latter and human cells is made basically through the paracellular pathway; therefore, the permeability of the colonic mucosa plays a crucial role in it. Some authors have shown important changes in this function associated to different health disorders. Electrical Bio Impedance (EBI) techniques can be a useful tool for the study of the colon, with measurements taken from the luminal side. In this study, we report passive electrical response along the colon of rabbits (n = 7), where EBI measurements were carried out in 6 different segments. The results suggest that there are differences between the proximal and the distal portions of the colon. It is probably due to their differences in structure and functions, as it has been reported in other studies. Resistivity in the first part of the proximal colon seems to be higher than in most of the distal colon (8.7 and 6.6 Ohm m, respectively) with a gradual decreasing along the former and some fluctuation along the latter.**

Keywords: **Electrical Bio-impedance (EBI), Colon permeability, Rabbits.**

I. Introduction

Until very recently, the colon was a neglected organ, where little attention was given to its role in health and disease [1]. In the present century, though, as stated in the title of the article by these two authors, the colon has moved from "banal" to "brilliant". This is due to the fact that the complexity of its physiology has begun to be studied and better understood, as well as its role (especially in its interaction with the microbiota) not only in "local" diseases, but also in systemic health problems like metabolic [2], autoimmune [3] and even neurological diseases [4]. Some authors, for instance, call the enteric nervous system (ENS) the "second brain" [5], due to the complexity of cell types and integrated circuits and its relation to the rest of the body, including the "first brain".

Colonic homeostasis is mediated by complex interactions between food, xenobiotics, our own cells, the colonic microbiota (mainly bacteria) and the layer of mucous gel that covers the whole organ [1]. It acts both as its habitat and a barrier against them. The number of the latter exceeds about 10x times the number of cells in our own organism, and has, as a whole, as a metaorgan, more than 2 million genes (the microbiome) [6]. There is a permanent communication and interplay between that microbiota and our own organism, and this crosstalk is of paramount importance for our health status. The permeability of the colonic epithelium, a characteristic mainly controlled by the tight junctions, plays an important role in the functioning of this organ [7] [8]. The paracellular via plays a key role, as it is the pathway by which direct contact is established between that microbiota and the immune system of the host [8]. Some authors have shown how the permeability of the colonic epithelium is augmented in different clinical conditions [9] [10] and we think that Electrical Bio-Impedance (EBI) techniques can be used for the assessment of such permeability. Theoretically, there are three main ways of doing it: from the serosal side [10], transmurally [11] [12] and from the luminal side. The first one has been used in animals, *in vivo* and with implanted electrodes, while the second one has been used *ex vivo*, with samples of resected tissue. The possibility is, though, open for *in vivo* transendoscopical measurements of electrical bioimpedance from the luminal side of the colon [11], in a similar way as Gonzalez-Correa *et al* [13] did in the esophagus. These authors reported resistivity measurements in esophageal and gastric resected tissue from humans and rats, where readings were taken *ex vivo* from the luminal side of the wall. They also made some *in vivo* transendoscopic measurements in patients with Barrett`s esophagus. Both metaplastic intestinal and gastric epithelia are columnar and

F. Simini and P. Bertemes-Filho (eds.), *II Latin American Conference on Bioimpedance*,
IFMBE Proceedings 54,
DOI: 10.1007/978-981-287-928-8_8

show relatively low bioimpedance values at lower frequencies, when compared to esophageal tissue (around 5 and 12 Ohm m at 19.2 kHz, respectively). Finally, when studying the colon, it is necessary to take into consideration, that there are important anatomical and physiological differences between its proximal and its distal parts. Snipes *et al* [14], for instance, measured the electrical potential difference between the luminal side of the colon in rabbits and blood (reference electrode placed inside the ear vein), and found that it has high values in the proximal segments of the colon (above 40 mV) that fall to values around 15 mV in the distal segments, and, then, gradually increase to a maximum of 40 mV in the rectum. They speculate that these findings may result from "...regional differences in the transport and permeability properties of the mucosa". This article reports measurements taken *ex vivo* from the luminal side of resected colon from sacrificed rabbits. We chose this experimental model as it is a small animal widely used in experimentation and, in the future, it will allow us to use them *in vivo* transendoscopic impedance measurements. The main aim of this exploratory experimental study was, then, to determine whether or not there are differences in the electrical impedance between the proximal and distal colon of rabbits, using the above mentioned approach.

II - Material and methods

In our experiments, we used colon specimens excided from 7 female New Zealand rabbits used in another study approved by the Institutional Animal Ethical Committee of the same Institution. Ages were between 10 and 12 months, they were weighing between 9-10 kg and they were subjected to an experimental Achilles tendinitis. Resection was completed immediately after the animals were sacrificed with an anesthetic overdose of Xilazine 5 mg/kg plus Ketamine 30 mg/kg and measurements were carried out within one hour after it. These were taken on 6 points in the colon: at 10 (Point 1), 20 (Point 2) and 30 (Point 3) cms distally measured from the caecum, and 30 (Point 4), 20 (Point 5) and 10 (Point 6) cms orally measured from the anus (figure 1). Impedance readings at 19.2, 38.4, 76.8, 153.6 and 307.6 kHz were obtained using a Sheffield-Mark3-Multi-Frequency-Tissue-Impedance-Meter, in the same way as reported by Gonzalez-Correa *et al* [13]. The probe was a tetrapolar array of round gold electrodes (0.8 mm diameter), separated 0.4 mm from each other and forming a rectangle embedded in a circle with a diameter of 2.5 mm. A total of 210 data were obtained (7 rabbits x 6 points x 5 frequencies).

Statistical analysis. A dual way design was used, in which the dependent variable resistivity was expressed as Ohm m, and the independent variables were six different colon segments with 7 repetitions. We used the following statistics with the program STATGRAPHICS V5.1. multifactor ANOVA, Pearson correlation coefficient and Tukey`s HSD (Honest Significant Difference) test. We looked for differences between the readings at the six different segments and accepted a significant statistical difference when P was < 0.05.

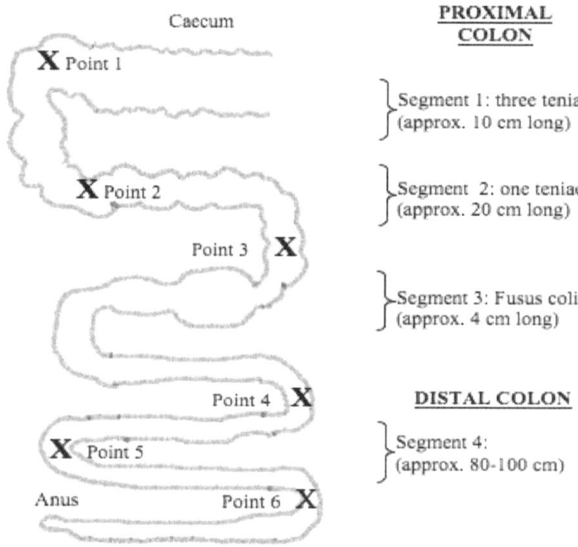

Figure 1. Anatomical placement of the 6 points where electrical impedance measurements were taken. Segments and approximate length estimations according to [14] and [15].

III - Results

Figure 2 shows the averaged resistivity, at 19.2 kHz, of the six points selected for this study. A two way ANOVA showed that the segment influenced the output variable ($P <$ 0.0000), and the Tukey`s test demonstrated that segment 1 had the highest average resistivity, followed by segment 2 and segment 3. Segment 4 showed a higher resistivity when compared to segments 3, 5 and 6, and was equal to the average of segment 2. The averaged values of segments 3, 5 and 6 were the same. In figure 2 we represent the values given by the Tukey test for the average of the resistivity discriminated by segments.

IV. Discussion

When resistivity is measured in the way we did, there are three main factors that can affect the readings: physical characteristics of the mucus layer (especially thickness and viscosity), status of the paracellular pathway and thickness of the colonic wall.

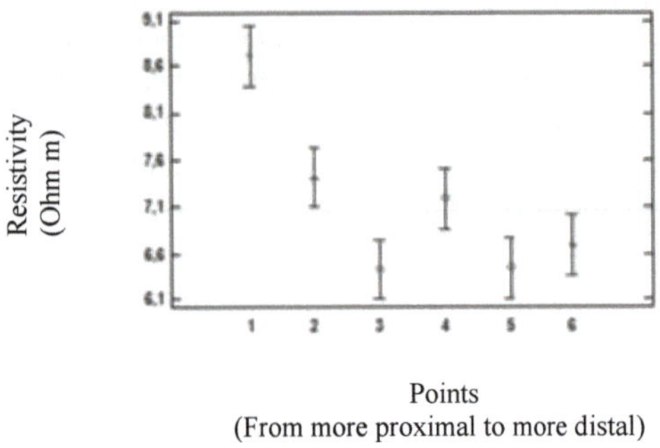

Figure 2. Tukey test for the averaged resistivity measured at 6 different segments on the colon of rabbits.

When two electrodes are used to inject a current in a homogeneous and isotropic medium, the penetration depth of the current is a function of its separation. In other words, the separation between electrodes determines the portion of tissue that can be measured by electrical impedance [16]. Given the very small separation between the two driving (current) electrodes, and considering the reported differences in the thickness of the mucus layer along the colon, we hypothesize that this fact can partially explain the differences in the resistivity found between proximal and distal colon (figure 2). In rats, for instance, thickness of the mucus layers in the colon can reach up to 800 µm [15] and the works by [17] [18] show that it increases distally from the caecum. In studies with monkies, it has also been found that thickness of the mucus layer increases from 10-80 µm in the proximal colon, up to 250 µm in the distal colon. In pigs, according to [17], there is an increase in thickness from the cecum to the ascending (proximal) colon, and a decrease from here to the descending (distal) colon, although they only took one measurement in each portion. As the distance between two adjacent electrodes in our system is only about 300 µm, as shown by [13] using Geselowitz theory [19], the penetration of the current can be estimated as approximately 500 µm. This all means that, to a good extent, the difference could be explained by the thickness of the mucus layer. In their article about modeling current distribution in cervical squamous tissue, for instance, [16] have shown how, even very thin layers of mucus on an epithelial surface, can have a large impact on the resistivity measurements. Another known fact that could have a role, though, is that the mucus viscosity increases progressively toward the distal colon, at least in the murine colon, a characteristic that determines a spatial distribution of bacteria [20]. However, it is also possible that differences in the proximal and distal anatomy and physiology can affect the electrical measurements. In the rabbit, for instance, and according to [14], the proximal colon is separated from the distal colon by the *fusus coli*, and has 2 different segments: a first one about 10 cm long and a second one of about 20 cm; the distal colon is between 80-100 cm long. The proximal colon in the rabbit is where the mix and absorption of water and electrolytes takes place, while propulsive forces for excretion are predominant in the distal colon [21]. Fermentative processes take place mainly in the proximal colon. In pathological situations, alterations in the tight junctions as well as edema in the colon wall could also influence the resistivity measured from the luminal side, but, probably, a wider separation of the electrodes would be desirable if we want to consider this aspect. Therefore, it would be interesting to make measurements in altered tissue to see if differences in electrical impedance can be found and can be associated with changes in the tissue. It is well known that, for instance, alteration in the tight junctions can cause or contribute to a wide variety of pathological conditions, including cancer [22] [23]. When edema in the intestinal wall is present, the electrical impedance varies as well [11]. Permeability is also altered when the integrity of the mucosa is lost [24] or when abnormal microbiota activate mucosal innate immune responses which increase epithelial permeability [25]. Finally, the results of our measurements, are similar to the profiles of the electrical potential difference (PD) in the colonic lumen to blood reported by [17], where we see a gradual decrease in the proximal colon, fluctuations in the distal colon and an increase at the end of the distal colon and rectum.

V - Conclusions

The data obtained in our experiments suggest that the colon mucosa has a relatively low electrical impedance with differences between its more proximal and its more distal parts. We found that electrical resistivity of colon seems to decrease gradually from proximal to distal. We hypothetize that these differences in electrical properties of the colon could also be observed in the human colon and it can be studied in order to determine abnormalities when colon wall thickness or the mucous layer change.

Compliance of ethical requirements

The animals used in our experiments were treated in accordance with the principles for care and use of experimental animals in the field of physiological sciences, with the approval of the Bio-Ethics Committee of the University of Caldas.

Acknowledgments

This study was partially supported by the University of Caldas.

CONFLICT OF INTEREST

The authors reported no proprietary or commercial interest in any product mentioned or concept discussed in this article.

REFERENCES

1. Sellers R, Morton D. The Colon: From Banal to Brilliant. (2013) Toxicol Pathol 42:67–81.
2. Karlsson F, Tremaroli V, Nielsen J, et al. (2013) Assessing the human gut microbiota in metabolic diseases. Diabetes 62:3341–3349.
3. McLean M, Dieguez Jr D, Miller L et al. (2014) Does the microbiota play a role in the pathogenesis of autoimmune diseases? Gut 64:332–341.
4. Mayer E, Knight R, Mazmanian S et al. (2014) Gut Microbes and the Brain: Paradigm Shift in Neuroscience. J Neurosci 34:15490–15496.
5. Avetisyan M, Schill E, Heuckeroth R. (2015) Building a second brain in the bowel. J Clin Invest 125:899-907.
6. Bermon S, Petriz B, Kajeniene A et al. (2015) The Microbiota: An exercise immunology perspective. Exerc Imunol Rev 21:70–79.
7. Camilleri M, Lasch K, Zhou W. (2012) Irritable Bowel Syndrome: Methods, Mechanisms, and Pathophysiology. The confluence of increased permeability, inflammation, and pain in irritable bowel syndrome. Am J Physiol Gastrointest Liver Physiol 303:G775–G785.
8. Wang X, Tully O, Ngo B et al. (2011) Epithelial tight junctional changes in colorectal cancer tissues. ScientificWorldJournal 11:826–841.
9. Peralta A, Miller R, Laughlin K et al. (1999) Increased tight junctional permeability is associated with the development of colon cancer. Carcinogenesis 20:1425–1431.
10. Gitter A, Bendfeldt K, Schulzke J et al. (2000) Trans/paracellular, surface/crypt, and epithelial/subepithelial resistances of mammalian colonic epithelia. Pflügers Arch – Eur J Physiol 439:477–482.
11. Radhakrishnan R, Shah K, Xue H et al. (2007) Measurement of intestinal edema using an impedance analyzer circuit. J Surg Res 138:106–110.
12. Huang Y, Huang E, Cheng K. (2013) The correlation between extracellular resistance by electrical biopsy and the ratio of optical low staining area in irradiated intestinal tissues of rats. Biomed Eng ONLINE 12:23–35.
13. González-Correa C, Brown B, Smallwood R et al. (1999) Virtual Biopsies in Barrett's Esophagus Using an Impedance Probe. Ann N Y Acad Sci 873:313–321.
14. Snipes R, Clauss W, Weber A et al. (1982) Structural and functional differences in various divisions of the rabbit colon. Cell Tissue Res 225:331–346.
15. Atuma C, Strugala V, Allen A et al. (2001) The adherent gastrointestinal mucus gel layer: thickness and physical state in vivo. Am J Physiol Gastrointest Liver Physiol 280:G922–G929.
16. Walker D, Brown B, Smallwood R et al. (2002) Modelled current distribution in cervical squamous tissue. Physiol Meas 23:159–168.
17. Varum F, Veiga F, Sousa J et al. (2010) An investigation into the role of mucus thickness on mucoadhesion in the gastrointestinal tract of pig. Eur J Pharm Sci 40:335–341.
18. Swidsinski A, Sydora B, Doerffel Y et al. (2007) Viscosity gradient within the mucus layer determines the mucosal barrier function and the spatial organization of the intestinal microbiota. Inflamm Bowel Dis 13:963–970.
19. Geselowitz D. (1971) An Application of Electrocardiographic Lead Theory to Impedance Plethysmography. IEEE Trans Biomed 18:38–41.
20. Corazziari E. (1999) Intestinal mucus barrier in normal and inflamed colon. J Pediatr Gastroenterol Nutr 48 Suppl 2:S54–S55.
21. Dinning P, Costa M, Brookes S et al. (2012) Neurogenic and myogenic motor patterns of rabbit proximal, mid, and distal colon. Am J Physiol Gastrointest Liver Physiol 303:G83–G92.
22. Chiba H, Osanai M, Murata M et al. (2008) Transmembrane proteins of tight junctions. Biochim Biophys Acta 1778:588–600.
23. Ryan C, Atkins M, Mier J et al. (1995) Effects of malignancy and interleukin-2 infusion on gut macromolecular permeability. Crit Care Med 23:1801–1806.
24. Marcon R, Claudino R, Dutra R et al. (2013) Exacerbation of DSS-induced colitis in mice lacking kinin B_1 receptors through compensatory up-regulation of kinin B_2 receptors: the role of tight junctions and intestinal homeostasis. Brit J Pharmacol 168:389–402.
25. Simrén M, Barbara G, Flint H et al. (2013). Intestinal microbiota in functional bowel disorders: a Rome foundation report. Gut 62:159–176.

Author: Edelberto Mulett-Vásquez
Institute: Universidad de Caldas
Street: Calle 25 # 48 - 57
City: Manizales
Country: Colombia
Email: emulettv@une.net.co

Assessment of Systolic Heart Function byWavelet Analysis of the Impedance Cardiogram

R. Stepanov[1], S. Podtaev[1], A. Dumler[1;2], and S. Chugainov[1;2]

[1] Institute of Continuous Media Mechanics, Perm, Russia
[2] Department of Introduction on Internal Diseasis, Perm State Medical Academy, Perm, Russia

Abstract—

Impedance cardiography (ICG) can be used to obtain one of the key hemodynamic parameters - stroke volume (SV). The SV is proportional to the left ventricular ejection time (LVET) and the maximum value of the first derivative of the recorded impedance (E wave) during the given cycle. Traditional ICG technique does not enable unambiguous detection of the LVET time between onset of the aortic valve opening and closing process. Objective of this work is investigation the possibilities of wavelet analysis (WA) approach to determine the parameters of the SV, in particular LVET. We can define LVET as the scale corresponding to the E wave maximum on 2D wavelet representation of the ICG data. Wavelet estimation of the LVET is well correlated (0.8) with the time interval between first and second heart sound, defined with usage of phonocardiogram. The proposed approach demonstrates the ability of ICG-WA technique to adequate assessment of SV parameters, including LVET and can be used in clinical practice.

Keywords— **impedance cardiography, wavelet analysis, hemodynamic parameters**

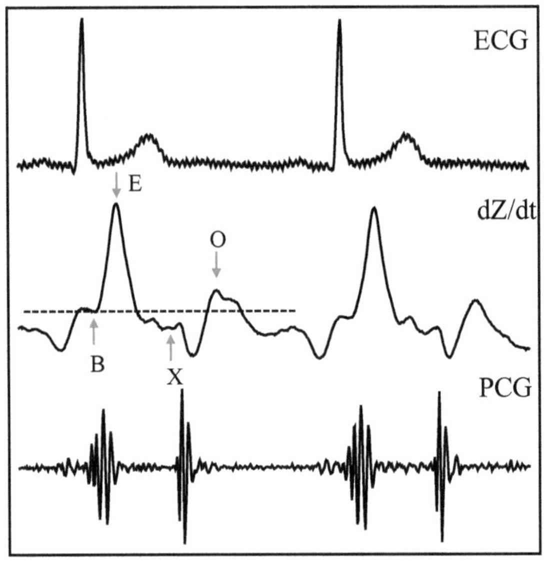

Fig. 1: Characteristic electrocardiogram (ECG), dZ/dt and phonocardiogram (PCG) signals, where B is the start of ejection of blood by the left ventricle, E is the major upward deflection occurring during systole, O is related to the diastolic phase, X the closure of the aortic valve.

I. Introduction

Impedance cardiography (ICG) is a simple, inexpensive, noninvasive technique for acquiring hemodynamic parameters. The impedance rheography method based on changes in the electrical resistance of a particular area of the body to high-frequency alternating current. These changes are proportional to variations in blood volume in the area at any given point in time. ICG can be used to obtain one of the key hemodynamic parameters: stroke volume (SV). In addition to the continuous hemodynamic monitoring of patients in intensive care, recognition of the different ICG patterns allows the rapid detection of cardiac dysfunction and the need for further cardiac evaluation. Compared with the standard ECG, the different patterns of ICG waveform are relatively easy to recognize and require considerably less time and skill to interpret than Doppler echocardiography [1].

There are several characteristic points which are usually considered to describe the ICG waveform [2]. These points are used to distinguish the physiological particularities of cardiac cycles. Figure 1 shows so called B, E ($(dZ/dt)_{max}$) X, O points.

The points B and X correspond to the first and second heart sounds. They are synchronous with the maximal deflection at the apex and the closure of the aortic valve, respectively. In some cases the identification of the B point location is difficult because the characteristic upstroke that provides a marker of this point is not always pronounced [3]. The E point is defined by the peak of dZ/dt curve which reflects the maximal derivative in the impedance. Ultrasound method measurements confirm that the E point may be associated with the maximal velocity of the heart ejection [4]. For ICG the first maximum (i.e. E-wave) is related to the systolic phase of the cardiac cycle and the second maximum with a smaller amplitude (O-wave) is related to the diastolic phase. The amplitude of the E-wave is proportional to the SV, and the am-

F. Simini and P. Bertemes-Filho (eds.), *II Latin American Conference on Bioimpedance*,
IFMBE Proceedings 54,
DOI: 10.1007/978-981-287-928-8_9

plitude of the O-wave correlates to a change in the volume of the left auricle during a short-term diastolic phase. In some cases the amplitude of the O-wave proves to be an important diagnostic parameter [5, 6]. Following the Kubicek formula [7], the stroke volume SV is proportional to the ejection time and the maximum value of the first derivative of the recorded impedance $Z(t)$ during the given cycle.

The left ventricular ejection time (LVET) is defined as the time interval of left ventricular ejection, which occurs between the opening of the aortic valve and its subsequent closure. However, as stated in some studies [8] there is evidence that ICG does not enable the detection of the onset of the aortic valve opening and closing process. Most of the ICG algorithms for LVET computation exhibit very low correlation coefficients and very high systematic estimation errors and dispersions [9]. Objective of this work is an demonstration of the wavelet transform for processing and interpretation of the impedance cardiogram waveforms and investigation the possibilities of this approach to determine the parameters of the SV, in particular LVET. For consistency we reproduce here some results of our previous study where the wavelet transform is suggested for an estimation of blood ejected velocity into the systole [10].

II. Measurements and its processing algorithm

We used the method of computer thoracic tetrapolar polyrheocardiography for simultaneous registration ECG, ICG and phonocardiogram (PCG) [11]. Participants group consists of eight healthy men of 20 to 25 years old and four patients, men of 50 to 55 years old, with diagnose of the essential hypertension. During polyrheocardiogram registration the functional test (isometric load) were performed for all participants. Patient's legs were held raised at an angle of 30-40 degrees to the horizontal in order to create a static force. Also all participants passed the thoracic tetrapolar polyrheocardiography without the functional test for comparison analysis.

Advanced signal processing techniques provides more elaborated approaches to the analysis and interpretation. In our study, we continue to promote the use of a continuous wavelet transform. We believe that its higher computational complexity are comparable to the Fourier transformation but the continuous wavelet transform makes the convenient assessment of amplitude, scale and phase of signal oscillations. Wavelet transform allows us to isolate a given structure in time and frequency space. Let us define the wavelet transform of the analyzing function $F(t)$ as

$$w_F(a,b) = \frac{1}{|a|}\int_{-\infty}^{\infty} F(t')\psi^*\left(\frac{t'-b}{a}\right)dt', \tag{1}$$

where $\psi(t)$ is the analyzing wavelet, a defines the scale (inverse to a frequency) and b defines the position in time of the wavelet. Then the coefficient w_F gives the contribution of corresponding structure into the function F.

The function F can be reconstructed using the inverse transform (see, e.g. [12])

$$F(t) = \frac{1}{C_\psi}\int_{-\infty}^{\infty}\int_{-\infty}^{\infty} \psi\left(\frac{t'-t}{a}\right) w_F(a,t')\,\frac{da\,dt'}{a^2}. \tag{2}$$

The reconstruction formula (2) exists under condition that

$$C_\psi = \frac{1}{2}\int_{-\infty}^{\infty}\frac{|\hat{\psi}(\omega)|^2}{|\omega|}d\omega < \infty. \tag{3}$$

Here $\hat{\psi}(\omega) = \int \psi(t)e^{-\imath\omega t}dt$ is the Fourier transform of the analyzing wavelet $\psi(\phi)$.

Let us define the wavelet transform of analysing function $G(t)$ which is derivation of measured function $F(t)$ as following

$$w_G(a,b) = \frac{1}{|a|}\int_{-\infty}^{\infty} G(t')\psi^*\left(\frac{t'-b}{a}\right)dt'. \tag{4}$$

Applying differentiation by part to equation (4) one can write

$$w_G(a,b) = \frac{1}{|a|}\int_{-\infty}^{\infty} F(t')\xi^*\left(\frac{t'-b}{a}\right)dt', \tag{5}$$

where $\xi(t) = -\psi'(t)$ is the differentiating wavelet. Then the wavelet coefficients of the derivative of the analyzing function $w_G(a,b)$ is obtained the wavelet transform with wavelet family $\xi(t)$. The advantage of original wavelet differentiation algorithm [13] is a combination of filtering and differentiating procedures to process the rheocardiogram. We use as the analyzing wavelet the so-called Mexican hat $\psi(t) = (1-t^2)\exp(-t^2/2)$ for higher resolution to separate characteristic points (see Fig. 1) in time. The wavelet Morle $\psi(t) = \exp(-\frac{t^2}{2} + 2i\pi t)$ were used for better spectral resolution.

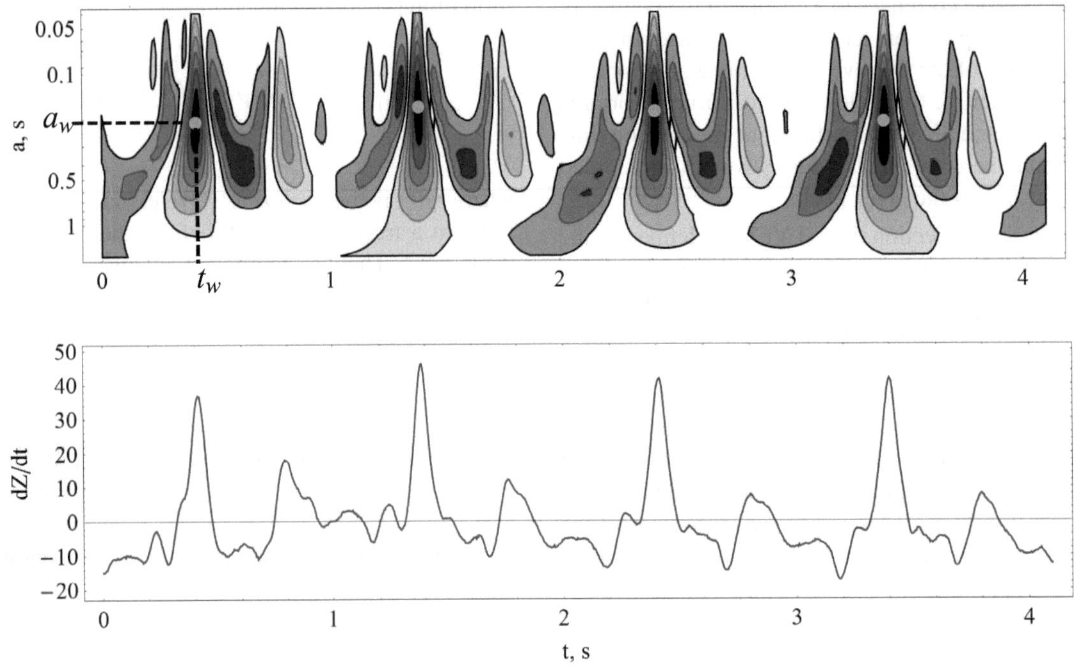

Fig. 2: 2D distribution of the wavelet coefficients (top) and analysing signal (bottom). The warm/cold colours denote positive/negatine values. White area corresponds to values close to zero. Characteristic E points are shown by pink points.

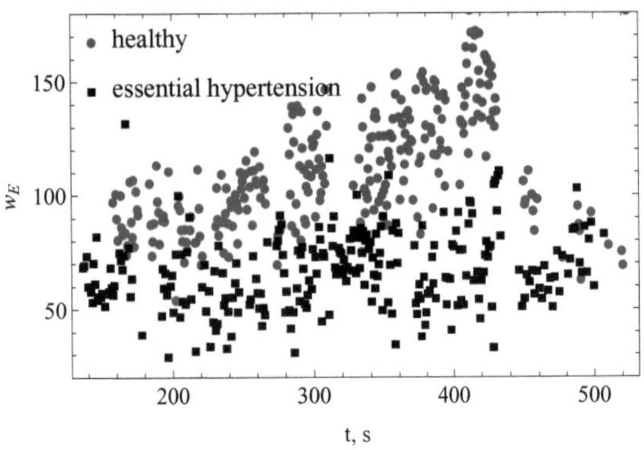

Fig. 3: Typical changes of the amplitudes E-wave waves after starting of the postural test (lifting leg at an angle of 30-40 degrees from horizontal until t=450 s) for healthy subject (red) and for the patient with hypertension (blue) (55 years old, hypertension in conjunction with coronary artery disease, with fixed dilation of the left atrium and concentric left ventricular hypertrophy).

III. RESULTS

Typical distribution of ICG wavelet coefficient (using the Mexican hat) is presented in Fig. 2. Principal difference between analysis of ordinal signal and its wavelet representation is that waves produce the local extremum in time and scale in 2D map. One can see positive defined (yellow) areas which correspond to E and O waves in each in the cardiac cycle.

It is possible to define time, scale and amplitude for each extremum point. We focus on the amplitude w_E and the scale a_w corresponding to the E wave maximum which can be defined for each cardiac circle. For verification conformity w_E and stroke volume we used hemodynamic response to isometric functional test (Fig. 3). In healthy subjects results characterized by a significant linear increase in stroke volume with increasing w_E. At the end of the load, indices have decreased to normal values during the first minute of recovery period. This response to the load is explained by the action of the Frank-Starling mechanism, which allows to implement an adequate hemodynamic response to stress. Changes in cardiac output were not found during isometric stress for for the patients with the essential hypertension. This means that the compensatory mechanisms of the immediate adaptation to hemodynamic stress is absent. All participant of each group show similar behaviour w_E as in Fig. 3.

The scale a_w of each extremum point with amplitude w_E characterizes an extension of E wave in time. It can be used for estimation the LVET. Let's consider the time interval between the first and second heart sounds as reference value. We defined it using the wavelet transform (using the Morle wavelet) of the phonogram which is recorded simultaneously with the impedance cardiogram. Then we calculate time interval between B and X points on the impedance cardiogram

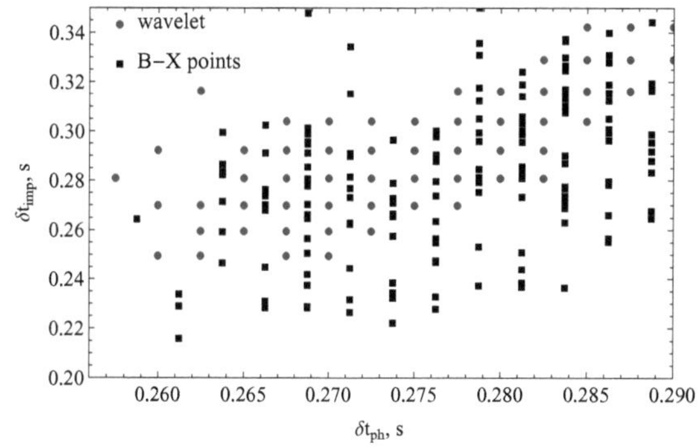

Fig. 4: Phonogram LVET vs impedance cardiogram LVET for selected healthy participant: δt_{ph} is defined as time interval between the first and second heart sounds, δt_{imp} is defined as B-X points interval (blue squares) or as the wavelet scale a_w of E-wave maximum (red points).

for each cardiac circle. The comparison of the LVET defined with usage of phonogram and impedance cardiogram for selected healthy participant is shown in Fig. 4. One can see that the wavelet estimate is well correlated with the time interval between first and second heart sound. The correlation coefficient calculated over 163 cardiac circles is 0.8. The classical estimate using B and X points agrees only with correlation coefficient equals 0.45. Similar correlations are found for the patient with hypertension. Averaged value correlation coefficients over all participants are 0.78 ± 0.07 for ICG-WA technics and 0.42 ± 0.12 for classical estimate using B and X points in ICG. We note that performing of the postural test is important to archive a significant variation of the heart rate. However the same condition can be provided by the deep breathing.

IV. Conclusions

The proposed approach demonstrates the ability of ICG-WA technics to adequate assessment of SV parameters, including LVET and can be used in clinical practice for early diagnostics of cardiovascular system remodelling in the course of different pathologies. We demonstrate an advantage of the wavelet transform not only as a common tool for filtration but also as an approach for introducing the new parameters defined by wavelet coefficients itself. Computational costs of the wavelet transform are comparable with ones for the fast fourier transform. So that required calculation are affordable using any laptops and the most of mobile gadgets, like phones and tablets.

Conflict of Interest

The authors declare that they have no conflict of interest.

Acknowledgements

The work is supported by the Russian Science Foundation under project 14-15-00809.

References

1. Bour Jean, Kellett John. Impedance cardiography - A rapid and cost-effective screening tool for cardiac disease *European journal of internal medicine.* 2008;19:399-405.
2. Lababidi Z., Ehmke D.A., Durnin R.E., Leaverton P.E., Lauer R.M.. The 1st derivative thoracic impedance cardiogram *Circulation.* 1970;41:651-658.
3. Lozano David L., Norman Greg, Knox Dayan, et al. Where to B in dZ/dt *Psychophysiology.* 2007;44:113-119.
4. Kerkkamp HJJ, Heethaar RM. A comparison of bioimpedance and echocardiography in measuring systolic heart function in cardiac patients in *Electrical bioimpedance methods: applications to medicine and biotechnology* (Riu, PJ and Rosell, J and Bragos, R and Casas, O . , ed.);873 of *Annals of the New York Academy of Sciences*:149-154Int Comm Promot Res Bio ImpedanceNew York Acad Sciences 1999. 10th International Conference on Electrical Bio-Impedance, Barcelona, Spain, apr 05-09, 1998.
5. Schieken R.M., Patel M.R., Falsetti H.L., Lauer R.M.. Effect Of Mitral Valvular Regurgitation On Trans-Thoracic Impedance Cardiogram *British Heart Journal.* 1981;45:166-172.
6. Pickett BR, Buell JC. Usefulness Of The Impedance Cardiogram To Reflect Left-Ventricular Diastolic Function *American Journal Of Cardiology.* 1993;71:1099-1103.
7. Kubicek W.G., Karnegis J.N., Patterson R.P., Witsoe D.A., Mattson R.H.. Development and evaluation of an impedance cardiac output system. *Aerospace medicine.* 1966;37:1208-1212.
8. Ermishkin V.V., Lukoshkova E.V., Bersenev E.Y., et al. Beat-by-beat changes in pre-ejection period during functional tests evaluated by impedance aortography: A step to a left ventricular contractility monitoring ;17 IFMBE:655-658 2007.
9. Carvalho P., Paiva R.P., Couceiro R., et al. Comparison of systolic time interval measurement modalities for portable devices :606-609 2010.
10. Podtaev S., Stepanov R., Dumler A., Chugainov S., Tziberkin K.. Wavelet analysis of the impedance cardiogram waveforms in *First Latin-American Conference On Bioimpedance (CLABIO 2012)*;407 of *Journal of Physics Conference Series* 2012.
11. Zubarev M, Dumler A, Shutov V, Popov N. Assessment of left ventricular systolic function and diastolic time intervals by the bioimpedance polyrheocardiographic system in *Electrical Bioimpedance Methods: Applications To Medicine And Biotechnology* (Riu, PJ and Rosell, J and Bragos, R and Casas, O . , ed.);873 of *Annals Of The New York Academy Of Sciences*:191-196New York Acad Sciences 1999.
12. Daubechies I.. Ten lectures on wavelets *Applied Mathematics.* 1992;61.
13. Stepanov R., Frick P., Shukurov A., Sokoloff D.. Wavelet tomography of the Galactic magnetic field. I. The method *Astronomy and Astrophysics.* 2002;391:361-368.

Author: Rodion Stepanov
Institute: Institute of Continuous Media Mechanics
Street: Ak. Korolev str. 1
City: Perm
Country: Russia
Email: rodion@icmm.ru

Impedance-Based Monitoring for Tissue Engineering Applications

C. Canali[1], A. Heiskanen[1], Ø.G. Martinsen[2,3], S. Mohanty[1], M. Dufva[1], A. Wolff[1], and J. Emnéus[1]

[1] Department of Micro- and Nanotechnology, Technical University of Denmark, Kgs. Lyngby, Denmark
[2] Department of Physics, University of Oslo, Oslo, Norway
[3] Department of Biomedical and Clinical Engineering, Oslo University Hospital, Oslo, Norway

Abstract— **Impedance is a promising technique for sensing the overall process of tissue engineering. Different electrode configurations can be used to characterize the scaffold that supports cell organization in terms of hydrogel polymerization and degree of porosity, monitoring cell loading, cell proliferation as well as the spatial distribution of cell aggregates in 3D. We have previously shown that impedance measurements allow accurate determination of conductivity in physiological solutions independent of validation and analysis of a specific equivalent circuit. Similarly to a physiological solution, cell culture medium conductivity, and hence the measured impedance, can respond to proliferating or dying cells populating the scaffold. Impedance may therefore be a key parameter for monitoring the biochemical dynamics that modulate 3D mammalian cell cultures over time. Furthermore, the conductivity of the medium filling the pores of the scaffold can serve as the basis for porosity determination using Archie's law. Different networks of structured or random channels and degree of porosity can be detected. In addition, by combining a number of two-, three- and four-terminal (2T, 3T, 4T) configurations, it is possible to obtain complementary information on spatial distribution of cells in a 3D scaffold. 2T- and 3T configurations also reflect the impedance at the interface between an electrode and cell-loaded scaffold (polarization impedance, Z_p), which may convey a further degree of information about the biochemical phenomena taking place in that sub-volume.**

Keywords—**Impedance, Electrode configurations, Scaffold porosity, Cell aggregate distribution in 3D cultures, Tissue engineering**

I. Introduction

Tissue engineering was first defined by Langer and Vacanti in 1993 as an interdisciplinary field that applies the principles of engineering (materials science and biomedical engineering) and life sciences (biochemistry, genetics, cell and molecular biology) to the development of biological substitutes which can reestablish, maintain, or enhance tissue functions [1]. Biological tissues can be fabricated in vitro by combining biomimetic 3D matrices ("scaffolds") of appropriate stiffness [2] and chemical composition, cells, and biologically active molecules (e.g. growth factors and hormones) [3]. Such 3D environments have been shown to better reflect the essential cellular functions and increase the predictive power of cell-based drug and toxicity screening [4]. However, a major technological challenge in tissue engineering is the non-invasive on-line monitoring of the 3D organizational complexity, comprising scaffold characterization, cell loading and proliferation. From the early studies on 2D cultures, a number of label-free techniques have emerged, capable of determining the cell-substrate interactions in a non-invasive and real-time manner. These include electrical impedance spectroscopy (EIS), quartz crystal microbalance, refractive index based techniques [5] and microscopic imaging [6]. The latter approach needs to be optimized for tissue engineering applications, especially when using large 3D scaffolds that do not allow deep light penetration. Additionally, the scaffold should be optically transparent and composed of materials with low autofluorescence [4, 7].

In 1984, Giæver and Keese proposed the use of EIS to monitor cell adhesion, spreading and proliferation on a flat surface by relying on the insulating properties of cell membrane, which affect the electrode/electrolyte interface (polarization impedance, Z_p) [8]. Such measurements opened up new perspectives for monitoring of, e.g., cell motility [9], wound-healing [10] and cytotoxicity [11]. However, tissue engineering and 3D culturing systems warrant further advances in impedance-based sensing to gather insights into scaffold architecture and porosity, spatial distribution of cells and chemical changes related to cell proliferation and cell-scaffold interactions. For such purposes, technical solutions can be adapted from physiological impedance measurements, which have been widely described by Grimnes and Martinsen and defined as bioimpedance [12]. In this context, miniaturization and the need for continuous long-term measurements make electrode position and stability, respectively, crucial factors influencing the measurements. Moreover, electrode number, spacing and orientation should be thoroughly optimized with respect to the geometry of the cell culture chamber to maximize the sensitivity field distribution [13, 14].

This paper gives an overview of different impedance-based methods that we have developed and could find application in the overall process of tissue engineering (Fig. 1), from scaffold characterization (e.g. hydrogel

F. Simini and P. Bertemes-Filho (eds.), *II Latin American Conference on Bioimpedance*,
IFMBE Proceedings 54,
DOI: 10.1007/978-981-287-928-8_10

polymerization and porosity) to cell loading, proliferation and 3D spatial distribution using different electrode configurations.

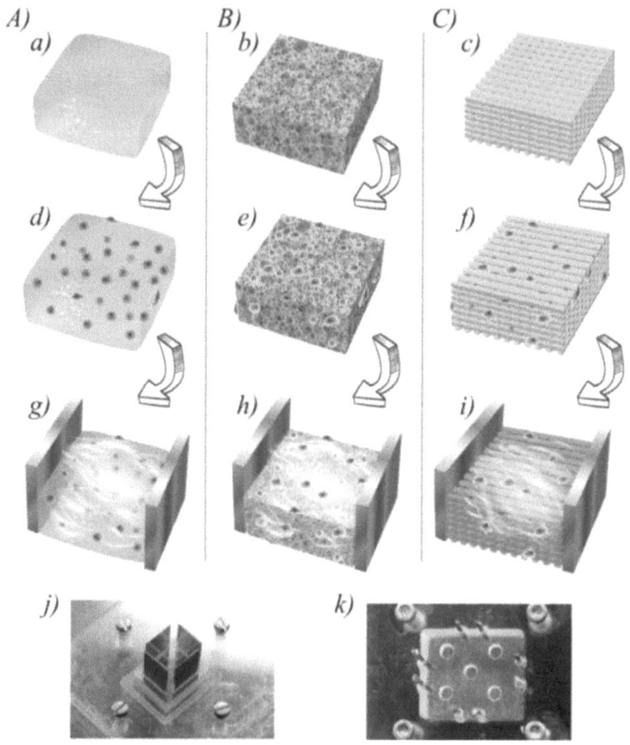

Fig. 1 Cell culture scaffolds can be characterized in terms of degree of polymerization (*A*), porosity (random pores, *B*, and structured pores, *C*), before (*a-c*) and after (*d-f*) cell loading using impedance (*g-i*) with different electrode configurations and setups (*j, k*) [13, 14].

II. MONITORING TISSUE ENGINEERING PROCESSES

In tissue engineering applications and 3D cell cultures, the scaffold properties (e.g. stiffness, chemical composition, porosity) and the supplied biochemical cues exert a major effect on cell physiology. Cell adhesion, proliferation and metabolism create an extremely heterogeneous and dynamic chemical environment [2, 3]. Variation in ionic strength leads to changes in conductivity, which is reflected by changes in the measured impedance. We have previously demonstrated a simplified approach for obtaining solution resistance (and, therefore, conductivity) by extracting the impedance magnitude at the frequency where the phase angle is closest to 0° without the need of validating and analyzing a specific equivalent circuit [15]. Additionally, changes in conductivity reflect hydrogel scaffold polymerization (e.g. gelatin, collagen), where mobile ions and readily reactive groups progressively lose their mobility depending on the degree of cross-linking (Fig. 1A) [13]. Moreover, we have shown that conductivity of the interstitial medium can be related to the scaffold's pore geometry by Archie's law (Fig. 1B,C) [16]. We optimized measurements using two gold plate electrodes interfaced with a scaffold (Fig. 1j) eliminating the necessity of penetrating the material, which provides a significant advantage of non-invasivity. Archie's empirical correlation basically corrects the bulk conductivity by a geometric factor, which depends on the cross-sectional area (*A*), length (*l*) as well as degree and type of porosity of the scaffold (Fig. 2). Furthermore, when impedance measurements are performed in protein-rich solutions, e.g., cell culture medium, the electrodes may require a protein-repellent functionalization in order to avoid biomolecule physisorption and allow reproducible measurements over time [15].

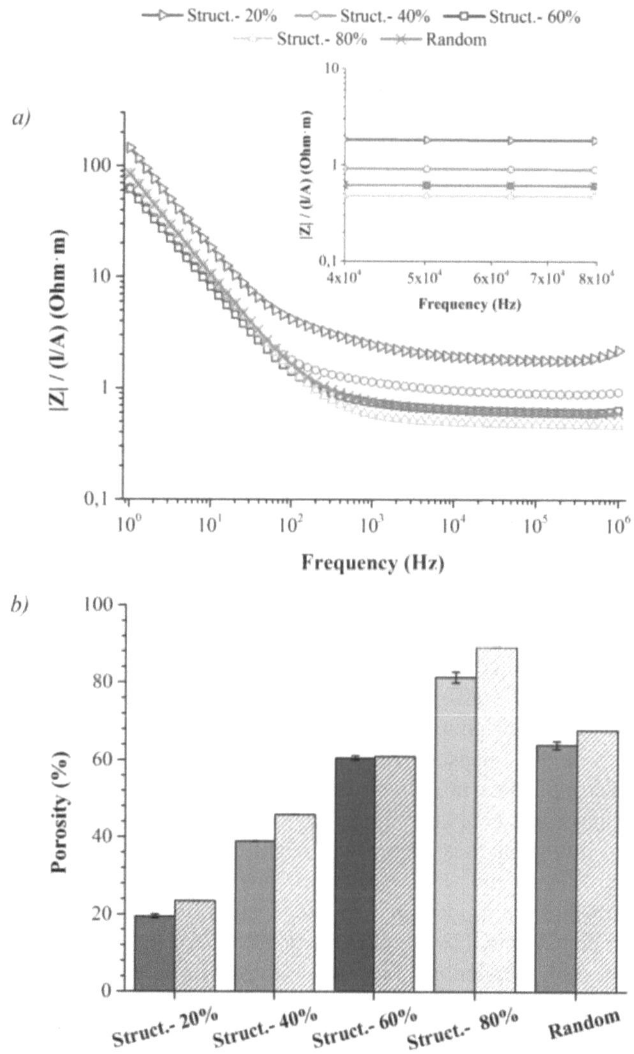

Fig. 2 *(a)* Normalization of the impedance data for porosity determination of 3D cell culture scaffolds displaying random or structured pores of decreasing size (i.e. 20 – 80 % of polymer composing the matrix). |Z| (Ohm) of the scaffold soaked in physiological buffer was divided by the geometric factor *l/A* (m^{-1}) characteristic of each scaffold. *(b)* Comparison of a more conventional weight-based method (■) and impedance measurements (▨). Data are reported as mean ± standard deviation (n = 3). [16]

A significant aspect in impedance monitoring of 3D cell cultures is that cells are not directly adhering on the electrode surface, but they are on a polymeric scaffold in contact with the electrodes via a conductive medium. Spatially distributed information on cell aggregates in the 3D environment can be collected using different combinations of electrode couples and configurations to maximize the sensitivity in the entire sample volume. The finite element method (FEM) is a valuable tool for tailoring parameters, such as electrode number, geometry, orientation and spacing, as well as predicting the distribution of the sensitivity field when using a specific electrode configuration [13]. Different 2-, 3-, and 4T configurations (Fig. 3) may be used in the same experimental setup to focus on specific sub-volumes inside the cell culture chamber that together cover the entire sample volume.

In the 2T configuration (Fig. 3*a*, *b*), the same electrode pair serves both as current-carrying (CC) and voltage pick-up (PU) electrodes and the measured biompedance typically includes the Z_p at both electrode surfaces, also reflecting the sample volume in close proximity to the electrodes. 2T measurements provide the simplest approach for conductivity analysis, especially when integrating bioimpedance monitoring in perfusion-based systems. In 3T configuration (Fig. 3*c*, *d*), one electrode is common for the CC and PU couple (working electrode, WE, in Fig. 3*c*). This is the only electrode inside the sensitivity zone of the measurement, and hence, the measured impedance is affected by its Z_p. We have previously presented the complementarity of 2T and 3T configurations in obtaining information on spatial distribution of cells in large gelatin scaffolds using plate electrodes [13].

In the 4T configuration (Fig. 3*e*, *f*), two separate CC and PU electrode couples are used. The CC couple is placed outside the measurement sensitivity zone, while the PU electrodes measure the so called "transimpedance". Such a concept is generally referred to as "bioimpedance" in physiological measurements. On virtue of modern voltage amplifiers, the PU electrodes are ideally non-current carrying, eliminating all Z_p contributions to the measurements. Hence, only the properties of the sample are detected. Sequential 4T measurements represent the best method to collect a large matrix of bioimpedance data from several electrode couples to map the impedivity of a sample, which also comprises changes in conductivity. By applying optimized algorithms, large data matrices provide possibility for constructing an image of the studied object, which is referred to as electrical impedance tomography (EIT, [17]). When using 4T measurements in tissue engineering applications, needle electrodes have a suitable geometry to achieve a high degree of 3D spatial information. We have evaluated the ability of 4T configuration using different CC and PU couple arrengements to sense the location of gelatin-embedded 3D cell aggregates [14].

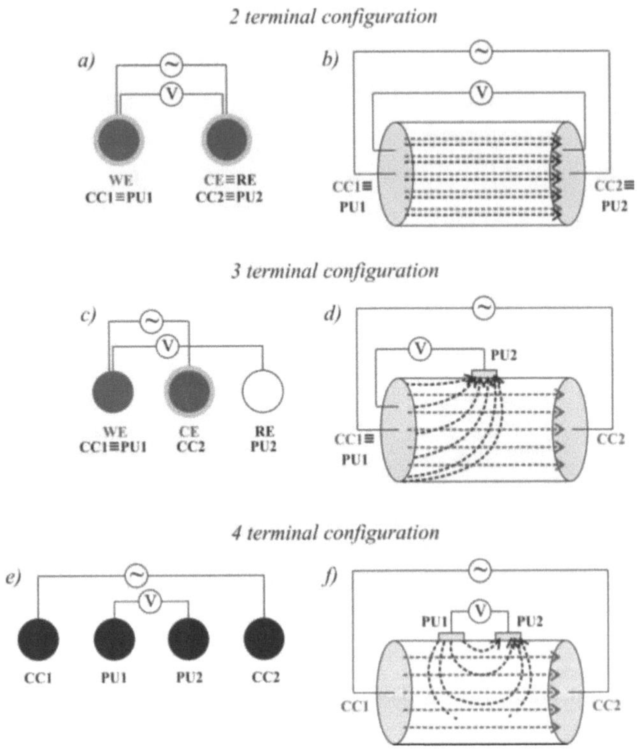

Fig. 3 Schematic of 2T (*a*), 3T (*c*) and 4T (*e*) configurations and their respective sensitivity fields (*b*, *d*, *f*) for a generic conductor. In electrochemical analysis WE (green) is the working electrode, CE (red) is the counter electrode, RE (white) is the reference electrode. In bioimpedance measurements CC1 and CC2 form the current-carrying couple; PU1 and PU2 form the voltage pick-up couple. When present, Z_p at the electrode/sample interface is reported in yellow (*a* and *c*). The sensitivity field (*b*, *d*, *f*) depends on the angle between the CC and PU equipotential lines (red and blue arrows, respectively) [18].

III. CONCLUSIONS

Bioimpedance-based methods can be applied in miniaturized 3D systems for future non-invasive monitoring of tissue engineering processes using different electrode geometries and configurations. The scaffold, supporting cell organization, can be characterized in terms of hydrogel polymerization and degree of porosity, as well as cell loading and proliferation. Information on spatial distribution of cell aggregates in a 3D scaffold can be obtained using the 4T configuration with multiple needle electrodes that map the impedivity distribution in the entire sample volume. Such a method paves the way towards image reconstruction of 3D cell cultures as an application of electrical impedance tomography in tissue engineering.

Acknowledgment

This study and the Ph.D. fellowship of C. Canali were financially supported by the EU-funded project NanoBio4Trans (grant no. 304842).

Conflict of Interest

The authors declare that they have no conflict of interest.

References

1. Langer, R., & Vacanti, J. P. (1993). Tissue engineering. *Science, 260*(5110), 920–926.
2. Engler, A. J., Sen, S., Sweeney, H. L., & Discher, D. E. (2006). Matrix elasticity directs stem cell lineage specification. *Cell, 126*(4), 677–689.
3. Griffith, L. G., & Swartz, M. A. (2006). Capturing complex 3D tissue physiology in vitro. *Nature reviews. Molecular cell biology, 7*, 211–224.
4. Pampaloni, F., Reynaud, E. G., & Stelzer, E. H. K. (2007). The third dimension bridges the gap between cell culture and live tissue. *Nature Reviews Molecular Cell Biology, 8*, 839–845.
5. Hug, T. S. (2003). Biophysical methods for monitoring cell-substrate interactions in drug discovery. *Assay and drug development technologies, 1*(3), 479–488.
6. Graf, B. W., & Boppart, S. A. (2010). Imaging and analysis of three dimensional cell culture models. *Methods in molecular biology, 591*, 211–227.
7. Smith, L. E., Smallwood, R., & Macneil, S. (2010). A comparison of imaging methodologies for 3D tissue engineering. *Microscopy research and technique, 73*(12), 1123–33.
8. Giæver, I., & Keese, C. R. (1984). Monitoring fibroblast behavior in tissue culture with an applied electric field. *Proceedings of the National Academy of Sciences of the United States of America, 81*(12), 3761–3764.
9. Giæver, I., & Keese, C. R. (1991). Micromotion of mammalian cells measured electrically. *Proceedings of the National Academy of Sciences of the United States of America, 88*, 7896–7900.
10. Keese, C. R., Wegener, J., Walker, S. R., & Giæver, I. (2004). Electrical wound-healing assay for cells in vitro. *Proceedings of the National Academy of Sciences of the United States of America, 101*(6), 1554–1559. doi:10.1073/pnas.0307588100
11. Caviglia, C., Zór, K., Montini, L., Tilli, V., Canepa, S., Melander, F., … Emnéus, J. (2015). Impedimetric toxicity assay in microfluidics using free and liposome-encapsulated anticancer drugs. *Analytical Chemistry, 87*, 2204–2212.
12. Martinsen, Ø. G., & Grimnes, S. (2014). *Bioimpedance and bioelectricity basics*. (Ø. G. Martinsen & G. Sverre, Eds.) (3rd ed.). London, UK: Academic Press.
13. Canali, C., Heiskanen, A., Muhammad, H. B., Høyum, P., Pettersen, F.-J., Hemmingsen, M., … Emnéus, J. (2015). Bioimpedance monitoring of 3D cell culturing – Complementary electrode configurations for enhanced spatial sensitivity. *Biosensors & bioelectronics, 63*, 72–79.
14. Canali, C., Mazzoni, C., Larsen, L. B., Heiskanen, A., Martinsen, Ø. G., Wolff, A., … Emnéus, J. (2015). An impedance method for spatial sensing of 3D cell constructs – towards applications in tissue engineering. *The Analyst* (in press). doi:10.1039/c5an00987a
15. Canali, C., Larsen, L. B., Martinsen, Ø. G., & Heiskanen, A. (2015). Conductometric analysis in bio-applications: a universal impedance spectroscopy-based approach using modified electrodes. *Sensors & Actuators: B. Chemical, 212*, 544–550.
16. Canali, C., Mohanty, S., Heiskanen, A., Muhammad, H. B., Martinsen, Ø. G., Dufva, M., … Emnéus, J. (2015). Impedance spectroscopic characterisation of porosity in 3D cell culture scaffolds with different channel networks. *Electroanalysis, 27*, 193–199.
17. Holder, D. (2005). Brief introduction to bioimpedance. In *Electrical impedance tomography* (1st ed., pp. 411–422). Bristol, UK: Institute of Physics Publishing.
18. Malmivuo, J. (2010). Principle of reciprocity solves the most important problems in bioimpedance and in general in bioelectromagnetism. In *Journal of Physics: Conference Series, 224*(1), 012001

Author: Chiara Canali
Institute: Technical University of Denmark
Street: Produktionstorvet Building 423
City: Kgs. Lyngby 2800
Country: Denmark
Email: chca@nanotech.dtu.dk

Total Body Water (TBW) for Body Composition Assessment in Young Adult Females from Colombia

Clara H. Gonzalez-Correa[1], Julio C. Caicedo-Eraso[2], and Dympna Gallagher[3]

[1] Research Group on Electrical Bioimpedance, Universidad de Caldas, Manizales, Colombia
[2] Department of Systems and Informatics, Universidad de Caldas, Manizales, Colombia
[3] New York Obesity Nutrition Research Center, St. Luke's–Roosevelt Hospital, New York, NY, USA

Abstract— **The study compared TBW derived from BIA, TBW derived from Watson (1980) anthropometrics formula with TBW from the gold standard, deuterium oxide dilution method in healthy Colombian young females. The purpose was to determine whether either or both of these clinic based measurement methods would allow for the accurate assessment of body composition. Subjects underwent multi-frequency BIA measurements, on the dominant side of the body, using an 800-µA and 50-kHz alternating current with a standardized tetrapolar technique. The BIA variables measured were R_{50} and Xc_{50}. The results show that TBW by anthropometry had a moderate coefficient of determination (r^2) and was not statistically different from TBW by D2O, however, the CI were wider than 2 standard deviations (SD). On the other hand, TBW by BIA had a better determination coefficient but was statistically different from TBW by D2O and also had wide CI. Both methods underestimated TBW. In conclusion, the present study suggests that BIA equations even when adapted to the specific population for the estimation of TBW should be used with caution within a specific population in order to assess individual differences. Anthropometric equations alone should also be interpreted wisely along with the clinic.**

Keywords— **Total body water, deuterium oxide dilution, anthropometry, bioelectrical impedance analysis, young females, Colombia.**

I. INTRODUCTION

One of the most important compartments to estimate when assessing nutritional status and body composition is the amount of water in the body (ie, total body water-TBW) as it is one of the most abundant and is highly related to fat-free mass (FFM). TBW measurements are useful in monitoring and treating fluid and electrolyte imbalances in athletes [1], patients with hormonal therapies [2], renal or hypertensive patients [3]. In addition, TBW estimates can be used to calculate body composition, especially fat free mass. The technique for measuring total body water accurately is the isotope dilution. However, this technique is laborious and is not routinely used in clinical settings or in epidemiological studies. Thus, it is necessary to find cheaper and easier methods that can be applied in in the environments. Anthropometry is one of the first choices, followed by bioelectrical impedance analysis (BIA). Both methods require the use of equations in which variables such as weight, height and sex of the subject are used. The differences in results obtained using both methods are likely due to the equations used for BIA which have been developed with subjects of distinct races and with diverse characteristics [4]. The purpose of this study was to compare the results of TBW obtained by the traditional Watson formula [5] and BIA equations designed for healthy Colombian young females [6] with TBW obtained by deuterium oxide dilution (D_2O) to evaluate whether they were comparable in terms of validity and reliability. The broader intention was to identify a tool to facilitate the assessment of total body water and thus the evaluation of body composition in hospitalized or outpatients in low income countries.

II. MATERIALS AND METHODS

A. Subjects and ethics

The procedures were approved by the Bioethics Committee of the Faculty for Health Sciences of the Universidad de Caldas (Colombia) and the Institutional Review Board of the Institute for Health Sciences of St. Luke's-Roosevelt Hospital Center (United States of America). Previous to sign an informed consent form, the purpose and procedures of the study were explained to subjects, who completed medical and physical-activity questionnaires to verify their health status, inclusion and exclusion criteria. Body composition variations specific to gender, age, ethnicity and genetic influence which affect BIA measurements were minimized by using a carefully designed protocol [7] and selecting a homogeneous group of young adult females. Forty young adult females from Colombia were evaluated. The same subjects and the same measurement data were used in another study (unpublished data).

B. Total body water (TBW) by deuterium oxide dilution (D_2O)

D_2O method is considered a reference method for in the in vivo measurement of TBW [8]. TBW by D_2O volume was calculated by equation 1 dividing the dose by the net D_2O concentration in the samples, where Cpost and Cpre are the Post-dose and Pre-dose deuterium concentration. The accuracy for TBW measurement as previously reported is±1.0%. [9,10,11].

F. Simini and P. Bertemes-Filho (eds.), *II Latin American Conference on Bioimpedance*,
IFMBE Proceedings 54,
DOI: 10.1007/978-981-287-928-8_11

$$TBW = \frac{D_2O\ dose}{C_{post} - C_{pre}} \quad (1)$$

C. TBW by anthropometry (ANT)

Anthropometric measurements were performed by the same trained technician [12]. All subjects wore underwear, a disposable hospital gown and no shoes. They were weighed to the nearest 0.1 kg with an electronic scale PP2000 (Icob-Detecto, A&D Co, Japan). Height was measured to the nearest 1 mm with a wall stadiometer Heightronic-235 (Seca, Hamburg, Germany). Weight and height were measured twice, and a third measurement was taken if it was found a difference greater than 0.1 kg or 5 mm respectively were found. Finally, TBW by ANT volume was calculated by equation 2 [5], with TBW in L, height (S) in cm and weight (W) in kg.

$$\text{TBW by } D_2O = -2.097 + 0.1107(S) + 0.2466(W) \quad (2)$$

D. FFM and TBW by bioelectrical impedance analysis (BIA)

Laboratory environmental conditions were controlled. Temperature (21.7 ± 0.7 °C) and relative humidity (RH) (61.2 ± 4.2 %RH) were monitored by thermo-hygrometer 13307 (±0.1°C/±1%RH, DeltaTrak, France) and controlled by electric heater FMH416 and dehumidifier BMD100 (Bionaire, USA) to avoid its influence on the measurements. Atmospheric pressure (788.1 ± 0.9 mmHg) was registered by barometer K4 (±0.1 mmHg, Konustar, USA).

Whole body wrist-ankle BIA was measured by the same trained person using standardized procedures [7] with the body composition analyzer Hydra ECF/ICF 4200 to the nearest 0.01Ω, 0.01° and 10 mL (Xitron Technologies Inc, San Diego, CA). Arrangements of four electrodes 292-STE (Impedimed, USA) was used, following the manufacturer's recommendations. Resistance and reactance raw data at 50 kHz (R_{50} and Xc_{50}) were obtained by software Hydra-S-Data-Acquisition-Utility (version 1.0, Xitron Technologies Inc). Prior to measurements, hydration, subject position; limb abduction and nonuse of jewelry were verified. A blanket was used to maintain a homogeneous skin temperature. Subjects remained in a supine position on a non-conductive surface for 5 minutes and the measurements were made three times between minutes 6 to 10 on the dominant side of the body, at the end of an exhalation. The arms were comfortably separated from the body 15 degrees and the legs were comfortably separated about 45 degrees. Dorsal hand and anterior foot surfaces were cleaned with alcohol and dried with a paper towel. Four landmarks were marked for the placement of the electrodes in order to obtain reproducible measurements. First, the midline between prominent ends of radius and ulna of wrist; second, the midline of third metacarpal-phalangeal joint on dorsal hand surface; third, the midline between the medial and lateral malleolus of ankle; and fourth, the midline of third metatarsal-phalangeal joint on anterior surface of foot. The current was applied at distal electrodes and the voltage was registered at proximal electrodes. FFM was calculated by BIA equation 3 [6], where FFM in kg, W in kg, R_{50} and Xc_{50} in Ω and S in cm.

$$FFM = 0.240\,W + 0.022R_{50} + 0.118Xc_{50} + 0.939\frac{S^2}{R_{50}} \quad (3)$$

TBW by BIA volume was obtained by equation 4, with age and gender hydration factor for the FFM of 72.9% for human [9,13] where TBW in L and FFM in kg.

$$TBW\ by\ BIA = FFM * 72.9\% \quad (4)$$

E. Statistical analysis

Mean and standard deviation (SD) were used to evaluate the laboratory conditions and the subject demographic characteristics. Comparison of TBW by D_2O against TBW by ANT and against TBW by BIA were developed using 3 approaches: a) simple linear regression to evaluate association with the determination coefficient (r^2) and standard error of estimate (SEE), b) paired Student 2-sided t-test to evaluate significant differences (p<0.05) and c) the Bland and Altman [14, 15] method to evaluate mean and limits of agreement. The $r^2 \geq 0.60$ and SEE≤2.50 L were considered as an acceptable goodness-of-fit for association, error and agreement [16]. All analyses were performed using XLSTAT software (ver. 2015.2.01.16868, Addinsoft).

III. Results

Subject's characteristics are shown in Table 1. Table 2 shows the results of the comparison of TBW by deuterium oxide dilution (D_2O) against TBW by anthropometry (ANT) and TBW by bioelectrical impedance analysis (BIA) by three complementary methods.

Table1. Subject demographic characteristics (n=40)

Variables	Mean	SD
Age [years]	20.9	2.1
Weight [kg]	54.0	6.1
Height [cm]	156.8	4.1
BMI [kg/m^2]	22.0	2.3
TBW by D_2O [L]	28.5	2.4
TBW by ANT [L]	28.0	1.7
TBW by BIA [L]	27.6	2.2
R_{50} [Ω]	646.1	53.9
Xc_{50} [Ω]	74.9	8.5

Table 2. Comparison of TBW by deuterium oxide dilution (D_2O) impedance analysis (BIA) against TBW by anthropometry (ANT) and TBW by bioelectrical (n=40).

TBW by D_2O against	Linear regression				t-test	Bland & Altman	
	Reported author		Obtained		p_t^a	Mean	CI
	r^2	SEE	r^2	SEE			
TBW by ANT	0.73	3.60	0.46	1.79	0.071	-0.53	-3,99 ;2,94
TBW by BIA	0.72	1.80	0.67	1.41	0.0003	-0.89	-3,63 ;1,85

ANT: anthropometry, BIA: bioelectrical impedance analysis, BMI: body mass index, D2O: deuterium oxide dilution, R50: resistance at 50 kHz, SD: Standard deviation, TBW: total body water, Xc50: reactance at 50 kHz.

Figure 1 shows the Bland and Altman plots of total body water (TBW) by deuterium oxide dilution (D_2O) against: TBW by anthropometry (ANT) and TBW by bioelectrical impedance analysis (BIA).

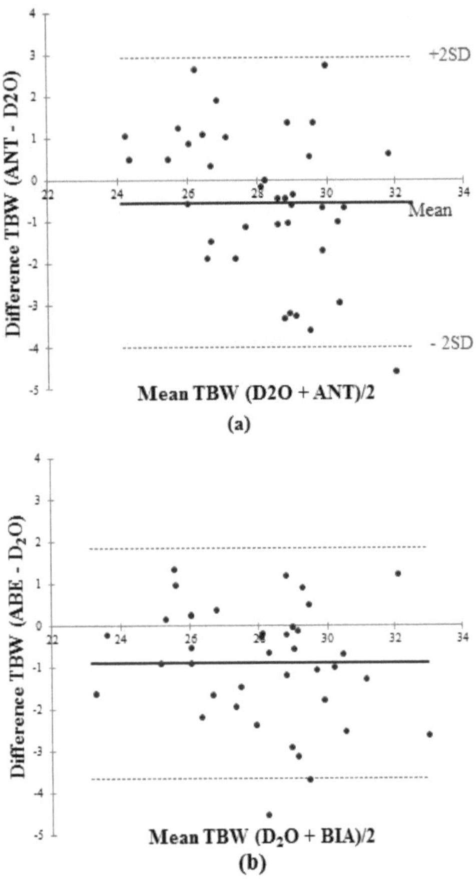

Figure 1. Bland and Altman plots of total body water (TBW) by deuterium oxide dilution (D_2O) against: (a) TBW by anthropometry (ANT) and (b) TBW by bioelectrical impedance analysis (BIA) (n=40).

IV. DISCUSSION

The evaluation of the interchangeability of two fast indirect measurements of TBW could facilitate clinical decision making in hospitalized and outpatient clinical settings. The results show that TBW by anthropometry had a moderate coefficient of determination and was not statistically different from TBW by D_2O, however, the CI were wider than 2 SD. On the other hand, TBW by BIA had a better determination coefficient but was statistically different from TBW by D_2O and also had wide CI. Both methods underestimated TBW. Other BIA equation [17] and anthropometric prediction equation [18] studies overestimated TBW compared to dilution techniques.

However, these researches included obese and elderly cohorts respectively in which body composition differ from the young people in this study [19].

Bioimpedance equation generated for our population, although had an acceptable coefficient of determination with respect to D_2O, and showed better agreement than reported in other studies, [18] had significant differences with D_2O. These results highlight the importance of the agreement test for interpreting results found using the coefficient of determination

The results are slightly more favorable for anthropometry than for BIA, since the percentage mean difference between anthropometry and dilution was almost the half of that with BIA (1.75 vs. 3.15%) and the interindividual variability in TBW was smaller than for BIA. However, this indicates that the estimation of the TBW volume calculated by anthropometry and bioimpedance are discordant with respect to that measured by D_2O, and both methods were not interchangeable. Similar results were found by de Fijter et al. [20].

BIA has many variables that can affect the results, especially the type of equation used. However, strengths of this study include an homogeneous sample in terms of age, weight, height, race and total TBW content as well as the strict measurement protocol followed and the use of an equation created specifically for the population examined here (2). Additionally, it was decided to use three parameters to evaluate the results.

The study has some limitations such as the sample size. Perhaps a larger sample size would improve and refine results. It would also be interesting to include patients with different clinical conditions where changes in TBW are prominent.

V. Conclusions

In conclusion, the present study shows that anthropometric and BIA estimates of TBW differed significantly from the gold standard and should not be used interchangeably to estimate TBW.

Acknowledgment

To Microbiology Laboratory staff at Universidad de Caldas for handling blood samples; Wenwen Yu and Body Composition Unit staff at New York Obesity Nutrition Research Center, St. Luke's–Roosevelt Hospital for D_2O analysis; and all volunteers for their cooperation.

Conflict of Interest

The authors declare that they have no conflict of interest.

References

1. Gonçalves EM, Matias CN, Santos DA, et al. (2015) Assessment of total body water and its compartments in elite judo athletes: comparison of bioelectrical impedance spectroscopy with dilution techniques. J Sports Sci.;33(6):634-40.
2. Mukherjee A, Adams Je, Smethurst L, et al. (2005) Interdependence of Lean Body Mass and Total Body Water, but not Quality of Life Measures, During Low Dose Gh Replacement in Gh-Deficient Adults. Eur J Endocrinol. 153(5):661-8.
3. Chen W, Cheng LT, Wang T (2007) Salt and fluid intake in the development of hypertension in peritoneal dialysis patients. Ren Fail; 29 (4):427-32.
4. Kumar S, Khosravi M, Massart A, et al. (2013) The Effects of Racial Differences on Body Composition and Total Body Water Measured by Multifrequency Bioelectrical Impedance Analysis Influence Delivered Kt/V Dialysis Dosing. Nephron Clin Pract; 124:60-66
5. Watson PE, Watson ID, Batt RD (1980). Total body water volumes for adult males and females estimated from simple anthropometric measurements. Am J Clin Nutr; 33(1):27-39
6. Caicedo-Eraso JC, Gonzalez-Correa CH, Gonzalez-Correa CA (2013). Preliminary bioelectrical impedance analysis (BIA) equation for body composition assessment in young females from Colombia. *J Phys Conf Ser*;434:012066. doi:10.1088/1742-6596/434/1/012066
7. Gonzalez-Correa CH, Caicedo-Eraso JC (2012). Bioelectrical impedance analysis (BIA): a proposal for standardization of the classical method in adults. *J Phys Conf Ser*;407:012018.
8. Ellis KJ (2001). Selected body composition methods can be used in field studies. J Nutr. 131(5):1589S-95S.
9. Schoeller, DA (2005). Hydrometry. In: Heymsfield SB, Lohman TG, Wang Z, Going SB, editors. Human Body Composition. Champaign, IL: Human Kinetics Publishers;. p. 35-49.
10. Yu W, Faruque O, Gallagher D, et al. (2005). An easy and inexpensive phantom for calibrating longitudinal water measurements by tracer dilution. FASEB J.; 19:A66
11. Gallagher D, Thornton JC, He Q et al. (2010). Quantitative magnetic resonance fat measurements in humans correlate with established methods but are biased. Obesity (Silver Spring). 18(10):2047-54. doi: 10.1038/oby.2010.97.
12. Lohman TG, Roche AF, Martorell R. (1988). Anthropometric standardization reference manual. Champaign, IL: Human Kinetics.
13. Keys A, Brozek J (1953).Body fat in adult man. Physiol Rev 33(3):245-325.
14. Bland JM, Altman DG (1986). Statistical methods for assessing agreement between two methods of clinical measurement. Lancet. 8;1(8476):307-10
15. Bland JM, Altman DG (1995). Comparing methods of measurement: why plotting difference against standard method is misleading. Lancet. 21;346(8982):1085-7
16. Baumgartner RN (1996). Electrical impedance and total body electrical conductivity. In: Roche AF, Heymsfield SB, Lohman TG (eds). Human Body Composition. Human Kinetics: Champaign, IL. pp 79–107
17. Martinoli R, Mohamed EI, Maiolo C et al. (2003) Total body water estimation using bioelectrical impedance: a meta-analysis of the data available in the literature. Acta Diabetol. 40 Suppl 1:S203-6.
18. Powers J S., Choi L, Bitting R, et al (2009) Rapid Measurement of Total Body Water to Facilitate Clinical Decision Making in Hospitalized Elderly Patients J Gerontol A Biol Sci Med Sci 64A(6): 664–669.
19. Johansson AC, Samuelsson O, Attman PO et al. (2001) Limitations in Anthropometric Calculations of Total Body Water in Patients on Peritoneal Dialysis. J Am Soc Nephrol 12: 568–573
20. de Fijter W M., de Fijter CWH., Oe PL, et al. (1997) Assessment of total body water and lean body mass from anthropometry, Watson formula, creatinine kinetics, and body electrical impedance compared with antipyrine kinetics in peritoneal dialysis patients. Nephrol Dial Transplant 12: 151–156

Corresponding author: Clara Helena Gonzalez-Correa
Institute: Universidad de Caldas
Street: Calle 65 # 26-10
City: Manizales
Country: Colombia
Email:clara.gonzalez@ucaldas.edu.co

Electrical Properties of Normal Cervical Human Cells in Suspension: The Relation between Normal Tissue and Electrical Impedance Spectrum

Lyda V. Herrera[1], C. Chaparro[1], Sandra P. Corzo[1], Sandra M. Pinto[1], G. Yáñez[1], Maryen Torres-Mejia[1], S.C. Méndez-Sanchez[1], and David A. Miranda[1]

[1] Universidad Industrial de Santander, Bucaramanga, Colombia

Abstract— **Epidemiological studies have shown that cervical cancer is a public health problem in several countries. Early detection of cervical cancer contributes with a better prognostic to the patient. Traditionally, Pap test has been used to screening cervical cancer and its early detection; however, this test has a low sensitivity and specificity. This work aims to study a new methodology to improve the early detection of premalignant lesions and cervical cancer studying electrical properties of cervical cells in suspension. A statistical analysis of Cole-Cole parameters extracted from electrical impedance spectrum (EIS) was used to study normal cervical cells in order to confirm if these parameters could represent normal cells. The results suggest that proposed methodology is appropriate to study cervical cells in suspension, but a study in abnormal cells is needed to validate the applicability to the early detection of cervical cancer using EIS.**

Keywords— **Electrical impedance, Cell suspension, Cole - Cole, Early detection, Cervical Cancer**

I. Introduction

Cancer itself is not a purely medical problem, but a complex socio-economic phenomenon and one of the most serious problems faced by humanity [1]. Cervical cancer is the second most common malignant tumor among women in the world. There are approximately 1.520.000 new cases of cervical cancer per year worldwide and approximately 780.000 deaths each year [2].

Although cervical cancer is invasive, its early detection leads to higher chance of cure. A regular screening reduces mortality of women due to cervical cancer [3].

There are many methods of cervical screening, among them those based on electrical properties of tissue are known by its high specificity and sensitivity [7].

Cancer detection by electrical properties of tissue is based on the physic-chemical principles of interaction between electric current and tissue. A useful parametric model to describe the variation of dielectric properties of tissues as a function of frequency is the proposed by Cole brothers [10, 11].

A method to study the behavior of cells is using a cell suspension; in this way the cells could be modeled as particles of the same shape but different size [7, 8].

In this work, we report a study of the electrical properties of normal cervical cells in suspension, in order to determine if the population of normal women can be represented by numerical values obtained by the electrical properties of tissue i.e. Cole-Cole parameters

II. Materials and methods

Endocervical and exocervical cells were collected from 33 women with ages from 20 to 60 years old. Two samples were taken from each patient and, two groups were conformed: One for Pap smear and, the second one for electrical measurements.

Figure 1. Bioimpedance measurements probe

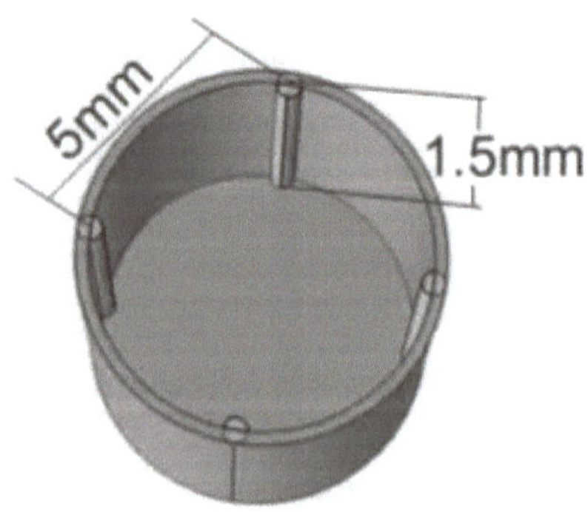

F. Simini and P. Bertemes-Filho (eds.), *II Latin American Conference on Bioimpedance*,
IFMBE Proceedings 54,
DOI: 10.1007/978-981-287-928-8_12

The second cell groups were placed into 1 mL of a phosphate buffered saline (PBS) solution (5 mM) and, a Neubauer chamber was used to estimate the cell concentration.

Serial dilutions were applied to initial exocervical suspensions cell in order to obtain a standard final concentration.

The electrical impedance spectra of initial and final cell suspensions were measured using a PGSTAT 204 AUTOLAB potentiostat and a tetrapolar probe (figure 1) with a 4μA sinusoidal current perturbation in the range from 1 Hz to 1 MHz.

Electrical impedance spectra were modeled with Cole-Cole model using the methodology described in [12]. A statistical analysis was performed to study the representation of normal cervix tissue by Cole-Cole parameters.

III. RESULTS AND DISCUSSION

A statistical analysis of the parameters obtained from the Cole-Cole inverse modelling of experimental data was performed using the algorithm proposed by Miranda et al. [12]. The one time-constant Cole-Cole dispersion model can be expressed by:

$$Z = Q_\infty + \frac{\Delta Q}{1+(j\omega\tau)^{1-\alpha}} [\Omega],$$

where, Q_∞ is the impedance at high frequency, ΔQ is the difference between the low and high frequency impedances $Q_0 - Q_\infty$, $j^2 = -1$, ω is the angular frequency, τ is the mean time constant and, $0 \leq \alpha \leq 1$ is a dimensionless parameter associated with the heterogeneity of the medium [12].

Our analysis was focused in the Q_0 parameter because of its relation with cervical abnormalities detected by electrical impedance spectroscopy in-vivo [15].

Table 1 shown the parameters extracted from experimental data represented in Figure 2 by the inversion of Cole-Cole model using Miranda et al methodology [12].

Table 1. Numerical values of the model parameters

Parameters	Q_0 (Ω)	Q_∞ (Ω)	ΔQ (Ω)	τ (s)	α
Values	162	8.4	153	0.042	0.3

Figure 2. Fitting simulated data to Cole-Cole model.

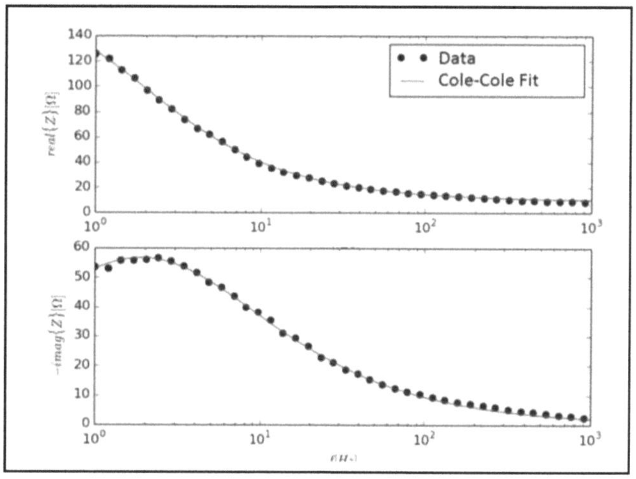

The statistical distribution of the parameter Q_0 was studied in order to associate this with normal stage in women. In this work only normal exocervical and endocervical cells were considered, i.e. based on the Pap test results. However, six specifics cases of the 33 initial were eliminated due to: parameters could not be obtained by a Cole-Cole. Then, in the statistical study were included experimental data for 27 women with negative Pap.

Table 2 shown basic statistics of each variable considered. The statistics values suggest that there are not reasons to deny normality of data.

Table 2. Descriptive statistics of Q_0 of the Initial and final variables.

Variable / Statistics	Initial Exocervix	Final Exocervix	Initial Endocervix
Mean (Standard desviation)	143 (65.94)	184.4 (68.02)	192.4 (95.07)
Median	142.5	182.5	182.5
Skewness	-0.205 (0.4)	0.029 (0.05)	0.381 (0.76)
Kurtosis	0.98 (0.93)	-0.956 (1.014)	-0.264 (0.918)
Minimum	18.7	74.8	39.7
Maximum	249.4	300.5	412.4

To substantiate, in the absence of evidence versus normality of variable, Table 3 shown the results test of skewness and, kurtosis besides Shapiro-Wilk test and, Shapiro-Francia test which are known by their validity in small samples. None test denies the normality of the variables, (p-value ≥ 0.33).

Table 3. Results associated tests to normality of data.

Variable	Test Normal data							
	Shapiro-Francia W'				Shapiro-Wilk			
	W'	V'	Z	Prob >Z	W	V	Z	Prob >Z
Initial exocervix	0.98	0.65	-0.79	0.78	0.97	0.89	-0.24	0.59
Final exocervix	0.98	0.57	-0.99	0.84	0.96	0.83	-0.36	0.64
Endocer-vix	0.97	0.83	-0.33	0.63	0.97	0.86	-0.32	0.62

Variable	Skewness/ Kurtosis Tests Normality			
	Pr (Skewnes)	Pr (Kurtosis)	Adj Chi2 (2)	Prob>Chi2
Initial exocervix	0.65	0.18	2.20	0.33
Final Exocervix	0.95	0.26	1.40	0.49
Endocer-vix	0.39	0.92	0.77	0.68

Because it was not possible to deny the assumption of normality of the data an ANOVA analysis was performed.

Tables 4 and 5 shown ANOVA results to three variables: initial exocervix, initial endocervix and, final exocervix, where the test does not present a statistically significant result to the level 5% (p-value:0,05). Similarly, Levene's test does not reject the equal variances.

Table 4. Anova Results.

Variables	ANOVA	
	f-Radio	*p-Value*
Undiluted Exocervix, undiluted endocervix and standard exocervix	**2.62**	**0.08**
	LEVENE	
	1.45	**0.24**

Table 5. Comparisons of ANOVA results for different cells in suspensions.

Comparisons	Means		Deviations	
	t-Ratio	p-Value	f-Ratio	p-Value
Undiluted Exocervix vs. Standard Exocervix	**-1.99**	**0.05**	**0.93**	**0.88**
Standard Exocervix vs. Undiluted Endocervix	**-0.31**	**0.76**	**0.51**	**0.15**
Undiluted Exocervix vs. Undiluted Endocervix	**-2.06**	**0.04**	**0.48**	**0.09**

In summary, the statistical analysis suggests that values of the parameters Q_0 in exocervix and endocervix of healthy women have the same normal distribution; therefore, any woman can be chosen to describe those values. That mean that cervical tissue could be represented by the numerical value of parameter Q_0. Based on previous results, this parameter takes significantly different values in each stage of neoplastic [15,16]. In this way and in accordance with Lopamudra et. al. results [17], we believe that with cells in suspension is possible to perform early detection of cancer in a similarly than with in-vivo measurements.

IV. Conclusions

We proposed a new methodology to measure electrical properties of cervical cells in suspension. A statistical analysis suggests that the parameter Q_0 of Cole-Cole model takes the same value for all women with normal cytology. Based on this result, we believe electrical properties of cells in suspension could be useful to screening cervical neoplastic.

Acknowledgment

The financial support of the Universidad Industrial de Santander (UIS), Bucaramanga Colombia (VIE-5742) and COLCIENCIAS is gratefully acknowledged.

Conflict of Interest

The authors declare that they have no conflict of interest.

References

1. S. Mukherjee. (2010) The Emperor of all Maladies: a biography of cancer. New York, pp 592.
2. J Ferlay, H Shin, F Bray et al. (2010) Estimates of worldwide burden of cancer in 2008: GLOBOCAN 2008. *International Journal Cancer*, vol. 127, pp. 2893-2917.
3. P. Sasieni, A. Castanon and J. Cuzick. (2009) Effectiveness of cervical screening with age: population based case-control study of prospectively recorded data. *British Medical Journal,* 339: b2968.
4. P. Cortiñas, I. Centeno, J. Sánchez, D. Martín (2009) Reunión de consenso en virus de papiloma humano 2008. Gaceta Médica de Caracas, vol. 117, pp. 49-69.
5. A. Waxman, ACOG Practice Bulletin. (2009) Cervical cytology screening. Clinical Management Guidelines for Obstetrician-Gynecologist. Number 109, pp. 12.
6. T. Wright Jr, L. Massad, C. Dunton, M. Spitzer et al. (2007) Consensus Guidelines for the Management of women with Cervical Intraepithelial Neoplasia or Adenocarcinoma in situ. In *American Society for*

Colposcopy and Cervical Pathology–sponsored Consensus Conference. American Journal Obstetrician- Gynecologist., vol. 197, pp. 346-355.

7. D. Dean, T. Ramanathan, D. Machado, R. Sundararajan. (2008) Electrical Impedance Spectroscopy Study of Biological Tissues. *Journal of Electrostatics.*vol.66(3-4), pp.165–177.
8. K. Asami, T Hanai, N Koizumi. (1980) Dielectric Approach to Suspensions of Ellipsoidal Particles Covered with a Shellin Particular Reference to Biological Cells. Japanese Journal Applied Physics, vol. 19, pp. 359-365.
9. H. Looyenga. (1965) Dielectric constants of heterogeneous mixtures, Physica, vol. 31, pp. 401-406.
10. K. Cole, H. Cole. (1941) Dispersion and absorption in dielectrics I. Alternating Current Characteristics. *Journal of Chemical Physics,* vol. 9, pp.341-351.
11. K. Cole, H. Cole. (1942) Dispersion and Absorption in Dielectrics II. Direct Current Characteristics. *Journal of Chemical Physics,* vol. 10, pp. 98–105.
12. D. Miranda, S. Jaimes, J. Bastidas. (2014) Assessment of carbon steel microbiologically induced corrosion by electrical impedance spectroscopy. *Journal Solid State Electrochemistry,* vol. 18, pp. 389-398.
13. A. Tarasov, K. Titov. (2013) On the use of the Cole–Cole equations in spectral induced polarization. Geophysical Journal International, vol. 195, pp. 352-356.
14. D. Miranda, S. Lopez-Rivera. (2008) Determination of Cole–Cole parameters using only the real part of electrical impedivity measurements. *Physiological Measurement,* vol. 29, pp. 669–683.
15. B. Brown, P. Milnes et al (2005) Detection of cervical intraepithelial neoplasia using impedance spectroscopy: a prospective study. International Journal of Obstetrics and Gynecology, vol. 112, pp. 802-806.
16. DA Miranda, S. C.-C. (2010). Early Detection of Cervical Intraepitelial Neoplasia in a Heterogeneos Group of Colombian Women Using Electrical Impedance Spectroscopy and the Miranda-López Algorithm. Journal of Physics: Conference Series , 407.
17. Lopamudra, D., Soumen, D., & Jyotirmoy, C. (2015). Electrical Bioimpedane Analisis: A New Method in Cervical Cancer Screening. Journal of Medical Engineering.

Author: David A Miranda
Institute: Universidad Industrial de Santander
Street: Carrera 27 Calle 9 Ciudad Universitaria
City: Bucaramanga
Country: Colombia
Email: dalemir@uis.edu.co

Analog Front-End for the Integrated Circuit AD5933 Used in Electrical Bioimpedance Measurements

F. Noveletto, P. Bertemes-Filho, and D. Dutra

State University of Santa Catarina (UDESC), Joinville, SC, Brazil

Abstract – **In the recent decades the bioimpedance has been used in many areas of medicine. The technological advance of electronics has enabled the increase of bioimpedance applications. Measuring biological tissues is a hard work, which leads to the need of continuous research based on better methods and equipments for diagnosis. Many electronics manufacturers have helped the biomedical area through the development of new electronic components and equipment. The objective of this work is to adapt the impedance meter AD5933 for bioimpedance measurements. Two Analog Front-End (AFE) circuits were built with different Voltage Controlled Current Source (VCCS) topologies, which are a load-in-the-loop and a mirrored modified Howland current sources. Circuits were implemented and both two and four electrode measurements in the frequency range of 5 to 100 kHz were performed. Results showed that both front-end circuits are suitable to be used with the AD5933 at lower frequencies. Also, it might be a low-cost alternative for electrical bioimpedance applications.**

Keywords – **Bioimpedance, Front-End Circuit, Current Source, AD5933.**

I. Introduction

The bioimpedance is defined as the ability that a biological tissue has to oppose the electric current passage [1]. These tissues react to the electrical current passage and can give an important information about the body health state. The bioelectrical impedance analysis (BIA) is widely used in body composition analysis. This technique evaluates the nutritional state of the body and can help in the diagnosis of diseases linked with the body fluids [2]. The BIA has been also benefited other medical areas such as cancer research, organ transplant and healthcare systems [3, 4, 5]. The biological impedance (Z) can be determined by injecting an alternate current (I) into tissue under study and measuring the voltage resultant (V). The result V/I is a complex impedance composed of a resistive part (R) and a capacitive one (Xc). R represents the opposition to the current flowing through the tissue and Xc represents the capacitive effect produced by cell membranes [1].

Although the technique is simple, hardware and software development for biological impedance measuring is a hard work. Many electrical bioimpedance (BIA) applications have been used dedicated circuits as, for example, the AD5933 (from Analog Devices), which is used as an alternative to reduce the complexity and cost of the system [6, 7, 8, 9]. This work develops two analog front-end circuits in order to adapt the impedance meter AD5933 for bioimpedance measurements.

II. Methodology

The main project requirements are the frequency control and the current amplitude to be injected into tissue, which is standardized by the IEC-60601 standard but it depends on the frequency. However, the AD5933 inject a sinusoidal voltage with constant amplitude in a bipolar configuration, which is not a recommended method for BIA human applications. Therefore, it was developed two different circuits in a four-electrode configuration. This approach reduces the influence of the contact impedance of the electrodes and is not load dependent. Both circuits use a VCCS circuit, where the injected current has constant amplitude and it is supposed to be load independent. The first Analog Front-End (AFE-1) circuit uses a Load-in-the-loop VCCS and the second one (AFE-2) uses a mirrored modified Howland VCCS.

A. Impedance Meter AD5933

The AD5933 is a high precision impedance meter consisted of a frequency generator (DDS), digital-to-analog converter (DAC), digital signal processor (DSP), analog-to-digital converter (ADC) and auxiliary circuits. It can generate a voltage signal with adjustable frequency up to 100 kHz with a resolution of 0.1 Hz. The current flow through the impedance is converted into a voltage and sampled by the ADC. The DSP processes both generated and received signal through an algorithm that performs Discrete Fourier Transform (DFT). The result is the real (R) and imaginary (I) part of the complex signal at each discrete frequency, which it can be accessed by the serial I2C interface. Both impedance magnitude and phase are calculated according to the component datasheet [10]. Figure 1 shows the AD5933 functional internal block and external connections, where Z(ω) is the biological load and RFB is the calibration resistor of 1 kΩ.

F. Simini and P. Bertemes-Filho (eds.), *II Latin American Conference on Bioimpedance*,
IFMBE Proceedings 54,
DOI: 10.1007/978-981-287-928-8_13

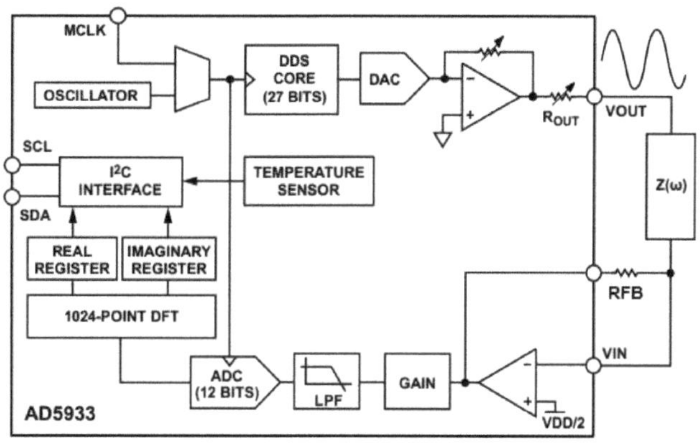

Fig. 1 AD5933 Functional Block Diagram [10]

B. Load-in-the-loop current source – AFE-1

This is a current source where the load (biological tissue) is connected in the negative feedback of the amplifier. A high pass filter (HPF) removes the DC component of the V_{out} signal (=1 Vpp). The current that flows in the load (1 mApp) is calculated by the ratio $V_{out}/R1$. The resistor R2 (=1 MΩ) is used to prevent the saturation of the Op-Amp and directly influences the output impedance of the source. An instrumentation amplifier (INA118) measures the voltage drop on the load, which is read by the AD5933. Due to the monopolar AD converter contained inside the AD5933, it is necessary an offset voltage (=Vdd/2) to be summed with the measured voltage.

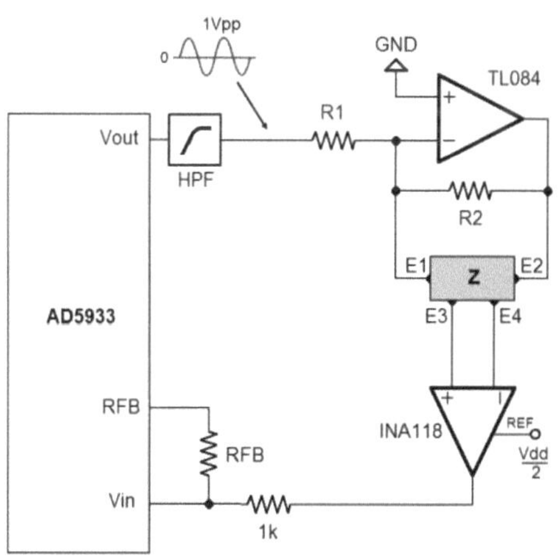

Fig. 2 Schematic of the Load-in-the-Loop Current Source

C. Mirrored Modified Howland Current Source – AFE-2

It consists of a VCCS composed by two symmetrical Howland-modified current sources connected to the load, as shown in Figure 3. The VCCS provides a constant output current (=1 mApp) that is given by the ratio between V_{out} (=1 Vpp) and the resistor *r* (=1 kΩ). All resistors have a low precision tolerance (±1%) in order to obtain a good current source performance [13]. In order to reduce power consumption, the resistor r of the Howland circuit (see Figure 3) should be much smaller than the value of R (=47 kΩ). The capacitor C (=0.1 μF) prevents DC offset voltages of the amplifier returning and then avoiding saturation. The circuit for measuring the voltage across the load is the same for both AFE-2 and AFE-1.

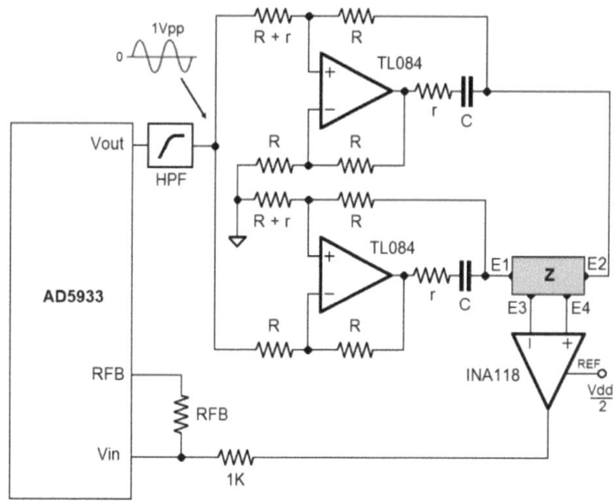

Fig. 3 Schematic of the Mirrored Howland-Modified Current Source

D. Microcontrolled System

A microcontroller (Atmega328 from Atmel) accesses the impedance data under test by means of the I2C interface from AD5933. The software in the microcontroller receives commands from the computer by the serial interface, which allows thc total control of thc AD5933. Thc softwarc permits the user to perform a frequency sweep with a user-defined start frequency, frequency resolution, and number of points in the sweep. In addition, the device allows the user to set both V_{out} and RFB values. The microcontroller sends the impedance data (both real and imaginary part) by its serial interface at each analyzed frequency. Data are sent in a text format.

III. RESULTS

The system was tested by using two typical biological impedance models. The calibration procedures were based on the AD5933 datasheet, accordingly [10]. In order to cover a wide range of calibration resistance values, the impedance were chosen accordingly. Resistors and capacitors have a tolerance of ±1% and ±10%, respectively. Figures 4 and 5 show the magnitude and phase of the impedances A and B, as well as their respective theoretical values in the frequency range of 5 to 100 kHz.

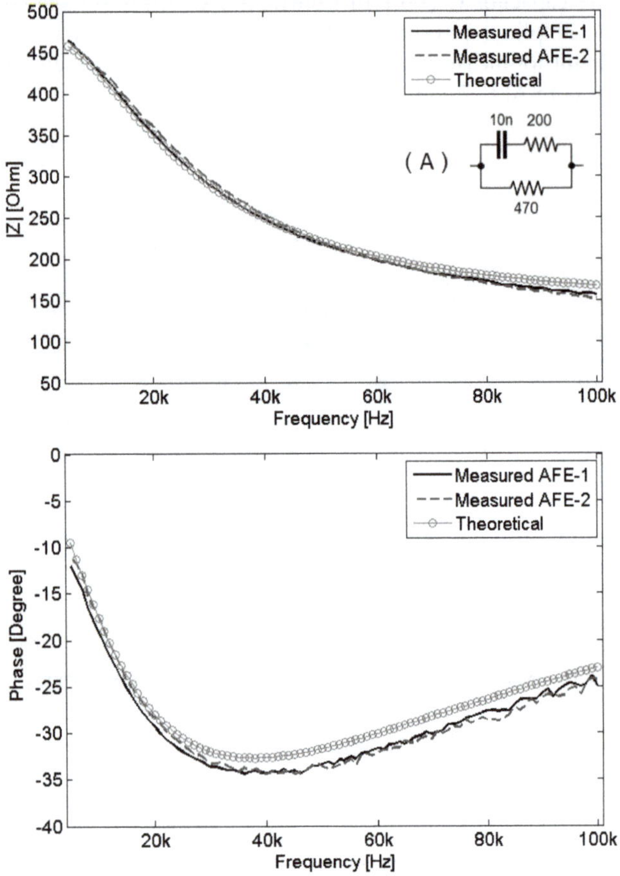

Fig. 4Frequency response of both module and phase load impedance A.

Both theoretical impedance module and phase (load A) range from 50 to 85% below the RFB value (=1 kΩ). On the other hand, it varies 20% above and 50% below the RFB. The AFE-2 presented a better performance than the AFE-1, especially for frequencies higher than 50 kHz. As expected, the errors increase with increasing frequency, but they were higher for the load B. Figure 6 shows the relative errors for both module and phase load impedance B. It was only analized at 5, 50 and 100 kHz, according to the literature of BIA systems.

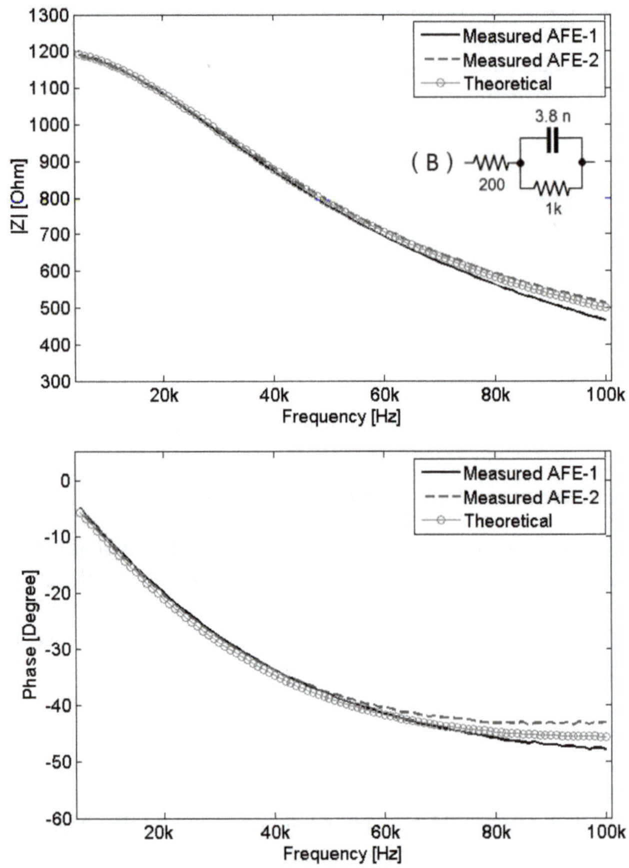

Fig. 5Frequency response of both module and phase load impedance B.

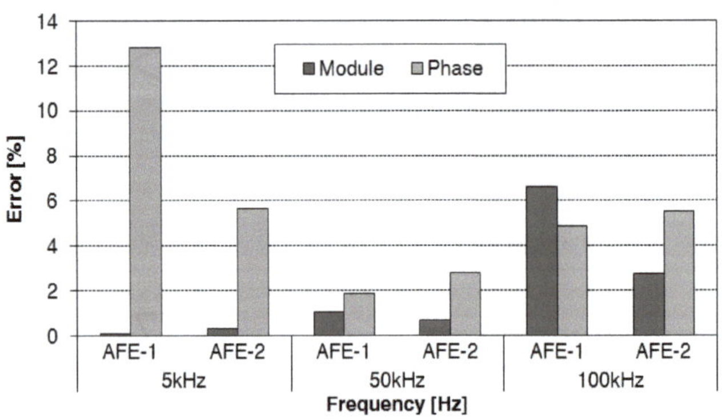

Fig. 6 Error analysis at 3 discrete frequencies for both module and phase load impedance B.

IV. Discussions

Topologies used in both current sources are simple and easy to be built. The topology shown in AFE-1 was widely used in the 60s, but it was replaced by other circuits due to its high frequency limitation required in many BIA applications. Recently, a research has been showed that the evolution of the electronic components enables the use of a simple VCCS topology (AFE-1) for BIA applications [11]. Acceptable results by Macias *et al* [12] were found in the AFE-1 topology when assessing the circuit performance up to 1 MHz. The AFE-2 topology has a good stability for the frequency range used in this work, but depends strongly on the symmetry between the current sources [13].

A critical limitation is the system calibration, where a resistive load has to be equal to the calibration resistor (RFB) when calculating the gain of the system [10]. Impedance values higher than 30% of RFB saturate the ADC input and compromises the system accuracy. Additionally, impedance values lower than RFB generates low voltages in the ADC input which is easily affect by noise interference. Thus, the correct choice of RFB is essential to determine the correct impedance load. However, in EBI applications the dynamic range is small and is possible to have a good precision for a proper calibration. The calibration problem is noted in the impedance A, where the values significantly lower than RFB (50 to 85%). In this case, the measured voltage is very low and can to explain the phase response slightly noisy. For the impedance B, the values range is relatively close to RFB and can be noted a similar value to the theoretical model. The obtained results were satisfactory at 50 kHz, which enable the AFE-1 and AFE-2 circuits to be used in a single frequency BIA analysis (SF-BIA).

V. Conclusion

The main requirements in this work were the low-cost and portability. Two different VCCS topologies were developed to adapt the AD5933 impedance analyzer for EBI using. It can be concluded that both proposed Front-End circuits are suitable for BIA applications in the frequency range of 5 to 100 kHz.

Aknowledgment

The authors thank the institutional and financial support of the State University of Santa Catarina (UDESC) and Technological Institute of Joinville (FITEJ).

Conflict of interest

The authors declare that they have no conflict of interest.

References

1. Martinsen ΦG and Grimnes S. (2008) Bioimpedance and Bioelectricity Basics. Academic Press, 2 Edition
2. Mialich MS, Sicchieriet JMF and Jordao-Junior AA. (2014) Analysis of Body Composition: A Critical Review of the Use of Bioelectrical Impedance Analysis. International Journal of Clinical Nutrition, Vol. 2, No.1, 1-10
3. Kamat DK, Chavan AP and Patil PM. (2014) Bio-Impedance Measurement System for Analysis of Skin Diseases. International Journal of Application or Innovation in Engineering & Management (IJAIEM), Volume 3, Issue 2, February 2014.
4. Fleury A, Sugar M and Chau T. (2015) E-textiles in Clinical Rehabilitation: A Scoping Review. Electronics 2015, 4, 173-203
5. Ferreira J, Seoane F, and Lindecrantz K. (2011) AD5933-Based Electrical Bioimpedance Spectrometer. Towards Textile-Enabled Applications. 33rd Annual International Conference of the IEEE EMBS, Boston, Massachusetts USA, August 30 - September 3, 2011
6. Pliquett U and Barthel A. (2012) Interfacing the AD5933 for bioimpedance measurements with front ends providing galvanostatic or potentiostatic excitation. First Latin-American Conference on Bioimpedance (CLABIO 2012), Journal of Physics: Conference Series 407
7. Martínez-Teruel J, García-Sánchez T, Fontova A et al. (2013) Electrical Impedance Spectroscopy cell monitoring in a miniaturized bioreactor. 119th IMEKO TC 4 Symposium and 17th IWADC Workshop Advances in Instrumentation and Sensors Interoperability, July 18-19, 2013, Barcelona, Spain
8. Margo C, Katrib J, Nadi M et al. (2013) A four-electrode low frequency impedance spectroscopy measurement system using the AD5933 measurement chip. Physiol. Meas. 34, 391–405
9. Seoane F, Ferreira J and Sanchéz JJ. (2008) An analog front-end enables electrical impedance spectroscopy system on-chip for biomedical applications. Physiol. Meas. 29, S267–S278
10. AD5933 Application notes at http://www.analog.com
11. Seoane F, Macías M, Bragós R et al. (2011) Simple voltage-controlled current source for wideband electrical bioimpedance spectroscopy: circuit dependences and limitations. Meas. Sci. Technol. 22, 11pp
12. Macías R, Seoane F and Bragós R. (2010) Performance of the Load-in-the-Loop Single Op-Amp Voltage Controlled Current Source from the Op-Amp Parameters. International Conference on Electrical Bioimpedance IOP Publishing Journal of Physics: Conference Series 224
13. Bertemes-Filho P, Felipe A and Vincence VC. (2013) High Accurate Howland Current Source: Output Constraints Analysis. Circuits and Systems, 4, 451-458

Author: Fabrício Noveletto
Institute: State University of Santa Catarina (UDESC)
Street: 200 Paulo Malschitzki
City: Joinville/SC
Country: Brazil
Email: fabricio.noveletto@udesc.br

Impedance Analysis for Medical and Electrochemical Applications Using a Low Cost Instrumentation

R.V. Tambara and L.O.S. Bulhões

Centro Universitário Franciscano, Santa Maria, Brazil

Abstract— **This paper presents a low cost instrumentation that is able to perform impedance analysis and the estimation of the electrode-electrolyte parameters through frequency sweep. The proposed system allows identifying the resistances and capacitance of the electrode-electrolyte equivalent model, where only voltage measurements are performed. Due to its flexible design, the system can be easily adapted to different applications and experimental requirements. Results show remarkable fair accuracy, considering the much lower price, smaller sizes and flexibility of use of the developed system.**

Keywords— **Electrode-electrolyte interface, Electrochemistry, Frequency response, Impedance analysis, Bioimpedance.**

I. INTRODUCTION

Impedance spectroscopy is a powerful technique to characterize electrochemical cells in equilibrium and during electron transfer reactions [1]. This technique is also used to the analysis of biological tissues with applications ranging from the analysis of the ratio of muscle mass and body fat [2] to the detection of bladder cancer [3]-[4] and breast cancer [5]. Impedance spectroscopy is a popular method of quantitative and qualitative monitoring of electrochemical process at electrode interface. The capacitance data is used in the study of adsorption of ions and neutral species onto metal surfaces [6]-[7] and is of use in measuring the change in double layer capacitance occurring during protein adsorption onto a surface [8]. In the last two decades the feasibility for integrated circuits makes the design and assembly of an impedance analyzer much simpler [9]. In order to simplify the use of impedance technique, devices such as oscilloscopes and function generators can be connected to a computer programmed using a graphical programming language that could managing the experiment. The application of a sine wave excitation to a system under test often is the easiest method of determining the system transfer function. In this paper, it is described a simple instrumentation to measure the impedance in electrochemical cell.

This paper is organized as follows: Section II presents the basic issues about modelling of the proposed system. Section III shows the proposed instrumentation structure. Section IV presents the experimental results of the proposed structure. Section V concludes the work.

II. MODELLING

A. Modelling of the electrode-electrolyte interface

The conventional electrical model of an electrochemical cell that represents the electrode-electrolyte interface (EEI) includes the association of resistances with capacitance as shown in Fig. 1. The parallel elements are related to the total current through the working electrode that is the sum of distinct contributions from the faradaic process and double-layer charging. The double layer capacitance resembles a pure capacitance, represented in the equivalent circuit by the element C and the faradaic process represented by a resistance, R_2. The parameters E and R_1 represent the equilibrium potential and the electrolyte resistance, respectively.

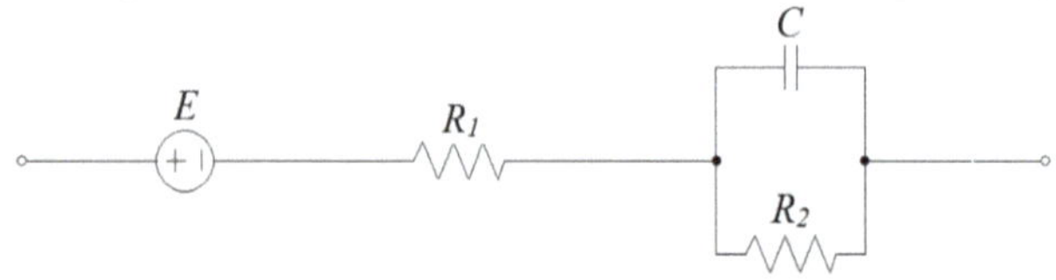

Fig. 1 Equivalent circuit of an electrochemical cell.

In order to obtain a systematic method to determine the EEI parameters using only voltage measurements, an additional resistor R_x is used. Figure 2 presents the respective equivalent electrical model plus a known value resistor R_x.

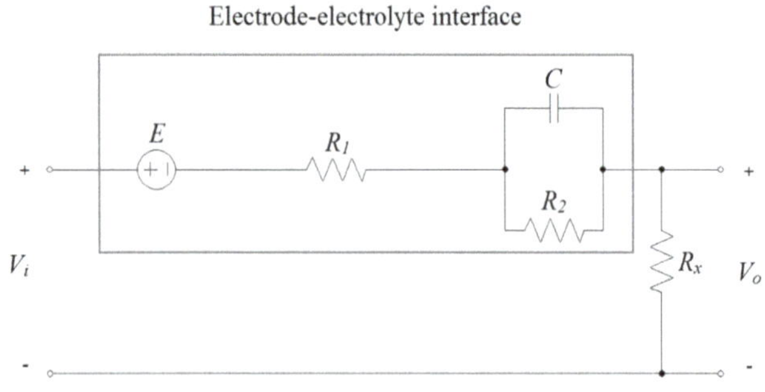

Fig. 2 Equivalent electrical model of EEI with a known value resistor R_x.

From Fig. 2, we can write the following transfer function

$$\frac{V_o(s)}{V_i(s)}=\frac{(R_xR_2C)s+R_x}{(R_xR_2C+R_1R_2C)s+(R_x+R_1+R_2)}. \quad (1)$$

F. Simini and P. Bertemes-Filho (eds.), *II Latin American Conference on Bioimpedance*,
IFMBE Proceedings 54,
DOI: 10.1007/978-981-287-928-8_14

The equation (1) has the zero

$$\overline{z} = -\frac{1}{R_2 C} \tag{2}$$

and the pole

$$\overline{p} = -\frac{R_x + R_1 + R_2}{R_x R_2 C + R_1 R_2 C}. \tag{3}$$

B. Estimation method of the parameters R_1, R_2 and C

By using frequency sweep, the parameters R_1, R_2 and C can be found. In sinusoidal steady-state, the Laplace variable $s = j\omega$ possibilities to rewrite (1) as

$$\frac{V_o(j\omega)}{V_i(j\omega)} = \frac{(R_x R_2 C)\, j\omega + R_x}{(R_x R_2 C + R_1 R_2 C)\, j\omega + (R_x + R_1 + R_2)}. \tag{4}$$

At low frequencies ($\omega \approx 0$), (4) can be expressed as

$$\left|\frac{V_o(j\omega)}{V_i(j\omega)}\right| \approx \frac{R_x}{R_x + R_1 + R_2}. \tag{5}$$

At high frequencies ($\omega \gg \left[|\overline{p}|, |\overline{z}|\right]$), (4) can be expressed as

$$\left|\frac{V_o(j\omega)}{V_i(j\omega)}\right| \approx \frac{R_x}{R_x + R_1}. \tag{6}$$

The capacitance C can be obtained by location of the zero in the gain diagram. By defining $\omega_z = |\overline{z}|$, from (2) is possible to write

$$C = \frac{1}{R_2 \omega_z}. \tag{7}$$

As $\omega_z = 2\pi f_z$ and f_z is the frequency of the zero (in Hz), (7) can be expressed as

$$C = \frac{1}{2\pi f_z R_2}. \tag{8}$$

C. Impedance estimation

In order to obtain the impedance modulus of the EEI, (4) can be written as follows

$$\frac{V_o(j\omega)}{V_i(j\omega)} = \frac{R_x}{R_x + Z(j\omega)}, \tag{9}$$

where $Z(j\omega)$ is the complex impedance of the EEI. Thus, from (9) is possible obtain the impedance modulus $|Z(j\omega)|$

$$|Z(j\omega)| = \frac{R_x \sqrt{1 + \left|\frac{V_o(j\omega)}{V_i(j\omega)}\right|^2}}{\left|\frac{V_o(j\omega)}{V_i(j\omega)}\right|}. \tag{10}$$

III. Instrumentation structure

An appropriated instrumentation is necessary to obtain the parameters correctly. Figure 3 presents the proposed system to identify the EEI parameters. This circuit has an input low-pass filter to attenuate the noise of the sinusoidal source V_{AC}. In this application, the peak value of the voltage source V_{AC} is 5 mV. Additionally, an output low-pass filter and amplifier is used to measure the noised and small voltage signal in resistor R_x. The voltage in R_x is amplified and the signal $V_o' = A_v V_o$ is obtained, where A_v is a known gain. Both input and output filter must have constant gain at interest frequencies (up to 10 kHz).

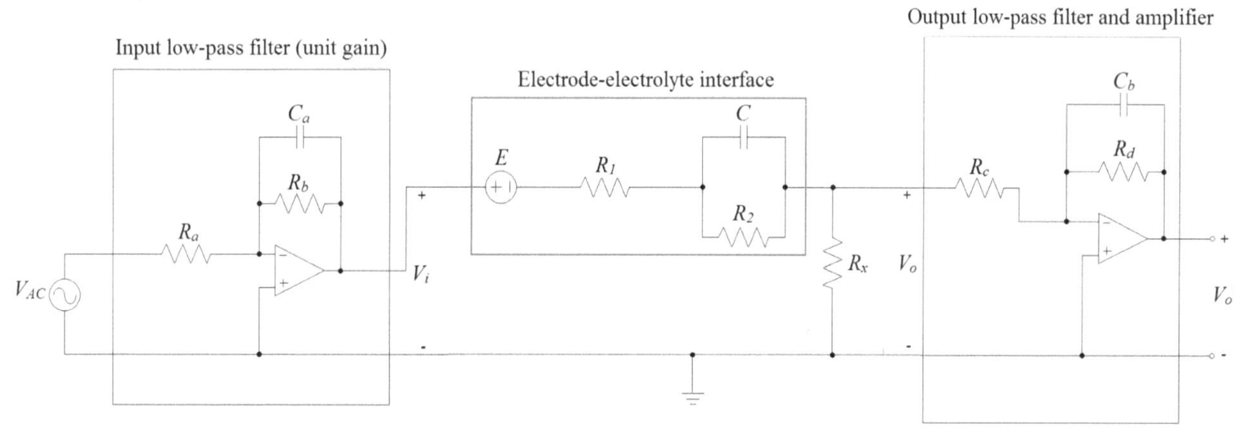

Fig. 3 Schematic diagram of the instrumentation used for impedance measurements.

A. Input filter design

The input filter is designed to provide unit gain and attenuation of high frequency noises. The transfer function is given by

$$\frac{V_i(s)}{V_{AC}(s)} = -\frac{R_b}{R_a}\frac{1}{(R_b C_a s + 1)}. \tag{11}$$

In order to obtain unit gain and cut frequency higher than 200 kHz, the following values are used

Table 1 Input low-pass filter elements

Element	Value
R_a	1 kΩ
R_b	1 kΩ
C_a	6.6 pF

B. Output filter design

The output filter is designed to provide high gain and attention of high frequency noise. The transfer function is given by

$$\frac{V_o'(s)}{V_o(s)} = -\frac{R_d}{R_c}\frac{1}{(R_d C_b s + 1)}. \tag{12}$$

In order to obtain high gain in a large frequency range (0.1 Hz up to 10 kHz) and cut frequency higher than 200 kHz, two configurations were used. Table 2 presents a list of components for an amplifier with a gain equal to 1000 in frequencies ranging from 0.1 Hz up to 500 Hz.

Table 2 Output low-pass filter and amplifier elements (from 0.1 Hz up to 500 Hz)

Element	Value
R_c	1 kΩ
R_d	1 MΩ
C_b	3.3 pF

Table 3 presents a list of components for the filter with a gain equal to 47 in frequencies ranging from 500 Hz up to 10 kHz.

Table 3 Output low-pass filter and amplifier elements (from 500 Hz up to 10 kHz)

Element	Value
R_c	1 kΩ
R_d	47 kΩ
C_b	3.3 pF

IV. EXPERIMENTAL RESULTS

A. Experimental Setup

For the electrochemical experiments was used an electrochemical cell with three electrodes, a platinum disc (0.2 cm^2 geometric area) as working electrode, a platinum flag (1.0 cm^2 geometric area) as auxiliary electrode and a Ag/AgCl in KCl 1 mol L^{-1} as reference electrode. The electrolyte was an aqueous Na_2SO_4 0.5 mol L^{-1} solution.

B. Results

Through a frequency sweep with a voltage source at 5 mV (peak value) and a fixed resistor R_x = 120 Ω, the gain analysis was carried out. Figure 4 presents the gain-frequency response of $|V_o(j\omega)/V_i(j\omega)|$.

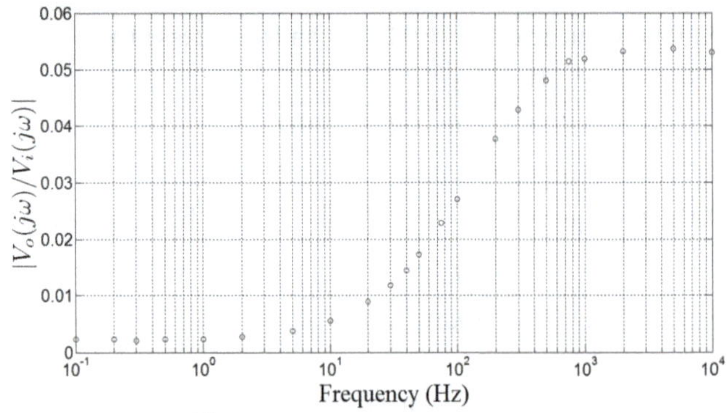

Fig. 4. Experimental gain diagram.

Figure 5 presents the gain-frequency response of $|V_o(j\omega)/V_i(j\omega)|$ in dB.

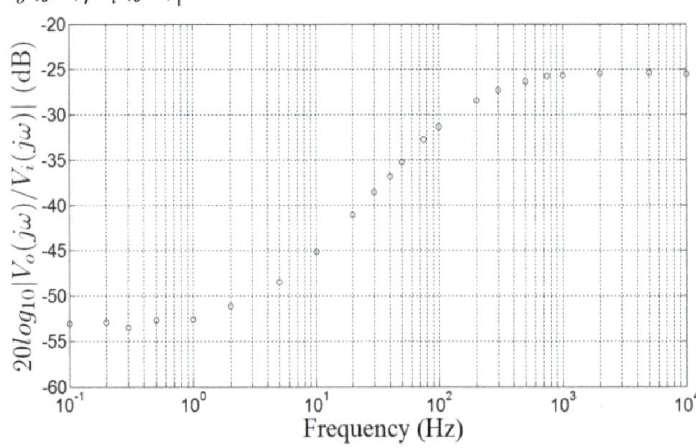

Fig. 5.Experimental gain diagram in dB.

By using (5) and (6) and Fig. 4, we can find the parameters R_1 and R_2. In high frequencies:

$|V_o(j\omega)/V_i(j\omega)| = 0.053$, then R_1 = 2.15 kΩ. In low frequencies: $|V_o(j\omega)/V_i(j\omega)| = 0.0022$, then R_2 = 51.5 kΩ.

By using (8) and Fig. 5, we can find the capacitance *C*. In Fig. 5, the frequency f_z (frequency of the zero) was estimated by interpolation: in this case $f_z = 3.26$ Hz (frequency where gain increase +3.01 dB), thus we can find *C* = 950 nF.

The equilibrium electrical potential *E* can be measured through a simple DC voltmeter.

Table 4 presents EEI parameters values.

Table 4 Experimental EEI parameters

Parameter	Value
E	0.37 V
R_1	2.15 kΩ
R_2	51.5 kΩ
C	950 nF

Finally, we can plot the impedance modulus $|Z(j\omega)|$ versus frequency using (10) and results of Fig. 4. Thus, Fig. 6 presents EEI impedance in interest frequencies.

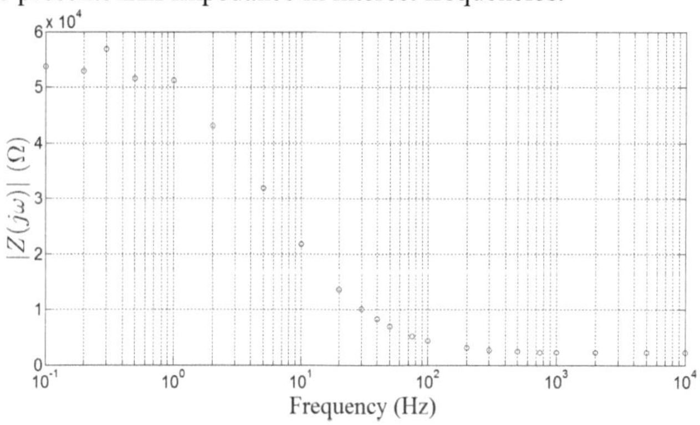

Fig. 6. Experimental impedance.

V. Conclusions

This paper have been presented a simple method to realize impedance analysis with estimation of the EEI parameters using a low cost instrumentation, where only voltage measurements are performed. The technique can be implemented using a computer and the technique can be applied to biological or electrochemical systems in continuous changing or a stable systems.

Acknowledgment

The authors wish to thank the Centro Universitário Franciscano and Brazilian Research Funding Institutions CNPq and CAPES for support.

Conflict of Interest

The authors have no conflict of interest to declare.

References

1. Barsoukov, E., J. Ross Macdonald, J. R. (Eds.), (2005), Impedance Spectroscopy: Theory, Experiment, and Applications, 2nd Ed Wiley, New Jersey.
2. De Lorenzo, A. Andreoli, A., Matthie , J., Withers, P., (1997) Predicting body cell mass with bioimpedance by using theoretical methods: a technological review, *J Appl Physiol* 82:1542-1558.
3. Li R., Gao J., A. Wang H., Jiang Q., (2013) Design of a Noninvasive Bladder Urinary Volume Monitoring System Based on Bio-Impedance, Engineeing, 5: 321-325.
4. Keshtkar A., Keshtkar A., Smallwood R. H., (2006) Electrical impedance spectroscopy and the diagnosis of bladder pathology. Physiol. Meas., 27:585-596.
5. Małecka-Massalska T., Chara K., Smolen A., Kurylcio A., Polkowski W., Lupa-Zatwarnicka K., (2012) Bioimpedance vector pattern in women with breast cancer detected by bioelectric impedance vector analysis. Preliminary observations, Annals of Agriculture and Environmental Medicine 19 (4): 697-700.
6. Lockett, V., Sedev, R., Ralston, J., Horne, M., Rodopoulos, T., (2008) Differential Capacitance of the Electrical Double Layer in Imidazolium-Based Ionic Liquids: Influence of Potential, Cation Size, and Temperature, J. Phys. Chem. C, 112 (190 7486-7495.
7. Wang, Z., Mu, X., Guo, M., Huang, Y., Mason, A. J., Zeng, X., (2013) Methane Recognition and Quantification by Differential Capacitance at the Hydrophobic Ionic Liquid-Electrified Metal Electrode Interface, Journal of The Electrochemical Society, 160 (6) B83-B89.
8. Farcas, M., Cosman, N. P., Ting, D. K., Roscoe, S. G., Omanovic, S., (2010) A comparative study of electrochemical techniques in investigating the adsorption behaviour of fibrinogen on platinum, Journal of Electroanalytical Chemistry, 649 (1-2) 206-218.
9. Jafari, H. M., Soleymani, L., Genov, R. (2012) 16-Channel CMOS Impedance Spectroscopy DNA Analyzer With Dual-Slope Multiplying ADCs, IEEE TRANSACTIONS ON BIOMEDICAL CIRCUITS AND SYSTEMS, 6 (5) 468-478.

Corresponding author:

Author: Rodrigo Varella Tambara
Institute: Centro Universitário Franciscano
Street: Rua Silva Jardim, nº 1295, Centro
City: Santa Maria
Country: Brazil
Email: rodrigo.tambara@unifra.br

Low-Cost Body Impedance Analyzer for Healthcare Applications

F. Noveletto[1], P. Bertemes Filho[1], D. Dutra[1], and A.V. Soares[2]

[1] State University of Santa Catarina, Depart. Electrical Engineering, Joinville, SC, Brazil
[2] Educational Association of Santa Catarina, Physiotherapy Clinic, Joinville, SC, Brazil

Abstract – **Bioimpedance has been widely used in many health areas. The ability of the human body cells, as the response for the alternate electric current flow, can provide valuable information about health status of the subject. This work presents a low-cost body impedance measuring system. The system is composed by the impedance meter AD5933, a four-electrode front-end circuit, a microcontroller system and a computer system for processing, visualization and data storage. The developed system calculates the resistance, reactance and phase angle (PA) from measured impedance spectra. Two subjects were assessed and the bioimpedance data was compared with data encountered in the literature. Predictive equations were used to calculate fat-free mass (FFM), fat mass percentage (%FM) and total body water (TBW). Preliminary results describe consistent with the reference values in the literature, indicating that the developed device is proper to body impedance measurement. The body impedance analyzer developed in this work was conducted according to the requirements of healthcare applications, such as portability and low cost. It can be concluded that the developed system can be used for body composition assessment over a wide aging range.**

Keywords – **Impedance Analyzer, AD5933, Body composition.**

I. Introduction

The human body is composed by about 60 to 100 trillion cells that are grouped by function to form tissues and organs [1]. The health status of the body depends on the cell functions. Most laboratorial imaging diagnosis applies invasive and expensive technologies in order to assess the state of the body health. Bioimpedance Analysis (BIA) has been widely used as a noninvasive and low cost alternative in many medical areas [2,3,4]. The cells react to the alternate current flow according to their health status. The technique consists of injecting an alternate current of low amplitude ($\leq$ 1 mApp), over a frequency range of tens of kHz, into the body and measures the resulting voltage. The voltage and current ratio (V/I) is a complex impedance (Z) composed by a resistance and a capacitive reactance part. The resistance (R) depends on the electric conducting characteristics of the body fluids. The capacitive reactance (Xc) is caused by the cell membranes, actuating as capacitors that change with the frequency. Figure 1 illustrates the current path through the cells and their respective electrical model. At lower frequencies, cell membranes block the current flow. At higher frequencies ($\geq$ 50 kHz), the current also flows through the cell membranes and then it causes a delay in the current phase due to capacitive effects.

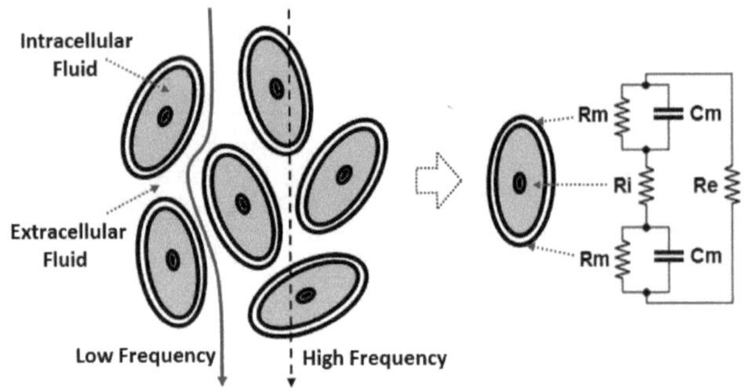

Fig. 1 Electrical current path through the cells, where Rm is the membrane resistance, Cm is the membrane capacitance, Ri is the intracellular resistance and Re is the extracellular resistance

Researches have associated BIA data with many diseases, which are caused by cells modification and then it changes the body impedance [5,6]. The phase angle is obtained from the geometric relationship between R and Xc, which has been used as an important indicator of cellular integrity. PA below 5% indicates some damage to the selective permeability of cell membranes. This characteristic is compatible with cell death caused in some types of cancer [7]. PA higher than 12% indicates large amounts of body mass and intact cell membranes, which is related to healthy subjects. It is shown in Figure 1 that, at low frequency, current does not flow through the cellular membranes, where extracellular fluid information is obtained. On the other hand, at higher frequencies, the current flows through the extracellular and intracellular fluids, allowing the determination of its reactance [2].

In order to better characterize tissue, a wide frequency range is required, as for example the body mass composition. Body mass is composed of high-conductivity tissues (lean body mass) and low conductive tissues (body fat). Excess body fat can lead to obesity and therefore increase the risk of serious health complications, such as heart disease and stroke [3]. BIA technologies used in body mass may vary according to the number of electrodes, signal excitation (voltage or current) and frequency range. Most

F. Simini and P. Bertemes-Filho (eds.), *II Latin American Conference on Bioimpedance*,
IFMBE Proceedings 54,
DOI: 10.1007/978-981-287-928-8_15

system use single-frequency excitation at 50 kHz (SF-BIA). Multi-frequency systems (MF-BIA) use few discrete frequency excitation ranging from 1 to 500 kHz in a four-electrode configuration [8].

The objective of this work is to develop a low-cost system for body impedance measurement in healthcare applications.

II. METHODOLOGY

In order to measure the body impedance, it was developed a homemade hardware and software. The system, shown in Figure 2, consists of the impedance meter AD5933 (from Analog Devices), an Analogue Front-End (AFE) circuit and a microcontroller for the computer interface.

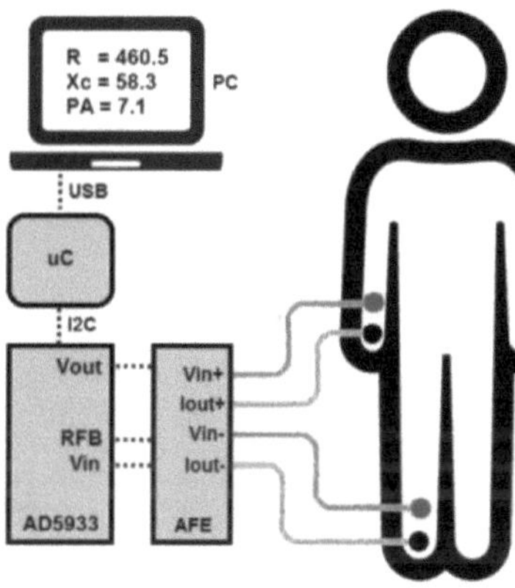

Fig. 2 Schematic of the body impedance analyzer, where PA is the phase angle, Iout+ is the inject current and Iout- is the return current.

A. System main characteristics

AD5933 – It is a high precision impedance meter, which includes frequency generator (from 1 to 100 kHz), digital-to-analog converter, analog-to-digital converter (12 bits) at 1 MSPS, digital signal processor (DSP) and auxiliary circuits [9]. The impedance data are processed by using a Discrete Fourier Transform (DFT) algorithm, which returns both real and imaginary part of the measured impedance at each selected frequency. Data are accessed by the I2C interface.

Analog Front-end – The output signal excitation of the AD5933 is a sinusoidal constant amplitude voltage with a DC level. It was manufactured to perform a two-electrode measurement. In order to use the AD5933 for human tissue measurements, it was used a four-electrode AFE circuit which complies according to safety regulations set by IEC-60601. A high pass filter removes the DC level of the reference voltage (V_{out}) to be converted into a current by a mirrored-modified Howland current source [10]. The output current is set to 800 µA over a frequency range of 5 to 100 kHz. The voltage across the load is measured by an instrumentation amplifier (INA118) and read by the AD5933.

Microcontroller and Computer Systems – The Arduino Nano board (from Atmel) is used as the interface between AD5933 (I2C) and the computer (USB). The embedded software in the microcontroller executes special commands to configure operating parameters of the AD5933 and data read are sent to the computer. The software allows the setting of some parameters of the AD5933, such as frequency sweep; number of collecting data, time between data collection, V_{out} amplitude and resistor calibration value. Impedance data are processed, converted in terms of R, Xc and PA values, which are stored in a text format.

B. Bioimpedance measurements

During the calibration process it is used a resistive load of 1 kΩ in order to calculate the system gain. Electrodes are placed in the right hemi-body of the subject and at the dorsal region of the foot and hand (see Figure 2). Current electrodes are connected at distal positions (near the toes) and the voltage electrodes are placed in the proximal positions away from the current electrodes at a distance of approximately 5 cm. The subject is rested in a supine position for 5 minutes before data acquisition. Also, the arms forms a 30° angle with the torso whereas the legs an angle of 45° between each other [8]. It was performed 5 measurements with an interval of 30 seconds between each other. Both mean R and Xc values are calculated in the frequency range from 5 to 100 kHz at 1 kHz intervals.

C. Body Composition

Both R and Xc values are used to calculate the body components, such as FFM, %FM and TBW. Two predictive equations from literature were used in this work, as shown in equations 1 and 2. The first one was obtained by using gold standard instruments, such as dual energy x-ray absorptiometry [11], which evaluated the bioimpedance and DXA data of 5225 healthy white subjects aging from 15 to 98 years old by determining the FFM (kg).

$$FFM = -4.104 + \frac{0.518 \times Height^2}{R_{50}} + 0.231 \times Weight + 0.130 \times Xc_{50} + 4.229 \times Sex \quad (1)$$

Where *Height* is expressed in centimeters, *Weight* in kilograms, both R_{50} and Xc_{50} values are at 50 kHz and *Sex* is 1 for male and zero for female.

The second prediction equation used a database from five North American research centers containing the body composition of 1774 white subjects aging from 12 to 94 years old by predicting the TBW [13].

$$TBW = 1.203 + \frac{0.449 \times Height^2}{R_{50}} + 0.176 \times Weight \quad (2)$$

Both results from prediction equations are compared to the measured data at 50 kHz by using the proposed BIA system.

III. Results

In order to validate the proposed BIA system, two healthy male white subjects were assessed. Table 1 shows the anthropometric characteristics of the subjects, as well as the measured impedance values and the literature ones based on the predictive equations 1 and 2.

Table 1 Anthropometric and bioimpedance characteristics at 50 kHz in comparison to the predictive equations 1 and 2 (Mean ± STD)

	Subject 1	Subject 2	Ref. [11]	Ref. [13]
Age (year)	45	47	45 – 54	40 – 49
Height (cm)	180	177	174.0 ± 6.4	178.9 ± 8.1
Weight (kg)	84.5	77.2	73.0 ± 9.2	82.2 ± 11.7
BMI (kg/m^2)	26.1	24.5	24.4 ± 2.6	25.7 ± 3.3
FFM (kg)	59.6	60.2	58.1 ± 5.1	61.7 ± 7.4
%FM	29.4	21.9	20.3 ± 5.1	24.3 ± 7.8
TBW (liter)	41.3	44.0	–	45.7 ± 6.7
R (Ω)	575.4	481.2	469.0 ± 43.0	–
Xc (Ω)	83.1	66.0	58.8 ± 8.4	–
PA (degree)	8.2	7.8	7.2 ± 0.9	–

Figures 3, 4 and 5 show the frequency response of the measured R, Xc and phase angle from two subjects, respectively. It can be observed in figure 3 that the spectra of the resistance R from subject 1 are higher than the subject 2, which might indicate a greater amount of fat mass. It can be observed in figure 4 that the reactance Xc represents approximately 14.4 and 13.7%, respectively, from subjects 1 and 2. In terms of phase angle from figure 5, it can be observed higher PA values in subject 2, which is compatible with BMI from literature [12].

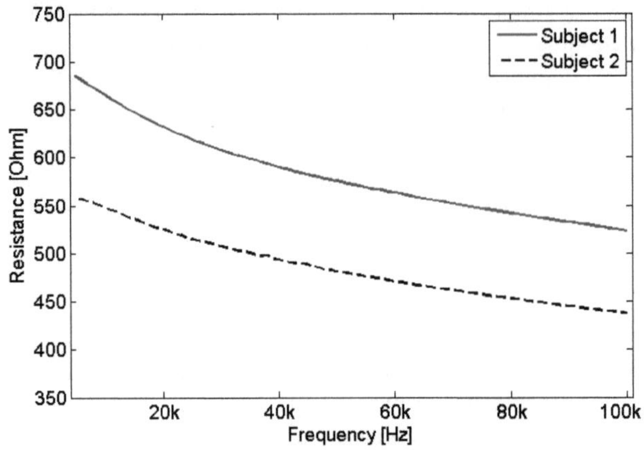

Fig. 3 Resistance R as a function of frequency from subjects 1 and 2

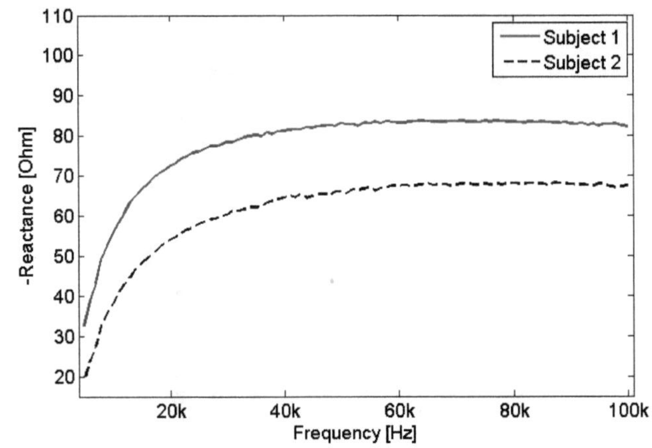

Fig. 4 Reactance Xc as a function of frequency from subjects 1 and 2

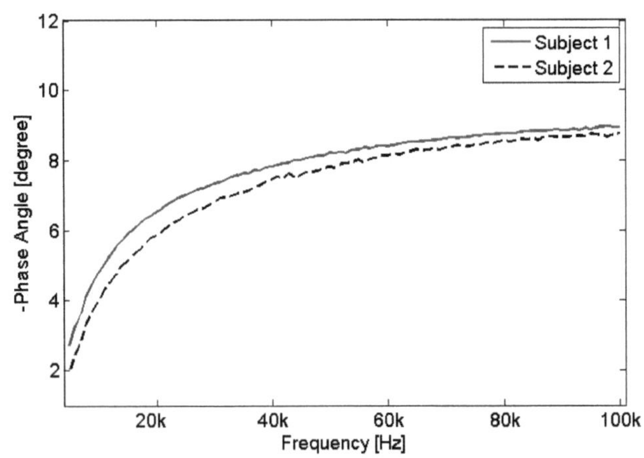

Fig. 5 Phase Angle as a function of frequency from subjects 1 and 2

IV. Discussions

BIA is a simple and low-cost technique for body composition diagnosis. The correct use of R and Xc with the predictive equations is important for calculating the body components, such as FFM and TBW. However, the estimation of the body composition by BIA depends on the collected data from DXA, for example. Although the predictive equations can change according to target population, there are many works in the literature that show equations with good correlation between bioimpedance and standardized methods [14].

The predictive equations used in this work presented small errors over the age range. FFM, %FM and TBW, based on measured R and Xc, were consistent with the ones reported in the literature [11,15]. In addition, other works have indicated that the phase angle is important for evaluating health status [7,16]. It is important to emphasize that PA value depends only on the R and Xc measurements and it is free from the statistical regression effects contained in the predictive equations. Barbosa-Silva *et al* (2005) determined PA values from 6.5 to 9.0 degrees of healthy adults aging from 40 to 49 years old. These findings are compatible with the results of this work. Caution has to be taken due to the fact that bioimpedance is a limited technique for body composition due to population differences. Even so, it is still considered an important tool for health professionals [8].

V. Conclusion

Bioimpedance is a non-invasive, safe and relatively inexpensive technique to assess body composition of both healthy and illness patients. The body impedance analyzer developed in this work was conducted according to the requirements of healthcare applications, such as portability and low cost. These characteristics are considered very important for large-scale use, benefiting the National Health System (NHS) and homecare users. It can be concluded that the developed system can be used for body composition assessment over a wide aging range.

Aknowledgment

The authors thank the institutional and financial support of the State University of Santa Catarina (UDESC) and Technological Institute of Joinville (FITEJ).

Conflict of interest

The authors declare that they have no conflict of interest.

References

1. Van De Graaff K M. (2003) Anatomia Humana. 6ª ed. Manole
2. Mulasi U et al. (2015) Bioimpedance at the bedside: current applications, limitations, and opportunities. Nutr Clin Pract, Apr; 30(2):180-93
3. Kyle UG et al. (2004) Bioelectrical impedance analysis - Part II: utilization in clinical practice Clinical Nutrition, 23, 1430-1453
4. Bayford R and Tizzard A. (2012) Bioimpedance imaging: an overview of potential clinical applications. Analyst, 137, 4635
5. Nescolarde L et al. (2015).Effects of muscle injury severity on localized bioimpedance measurements. Physiol. Meas. 36, 27-42
6. Kamat DK, Chavan AP and Patil PM. (2014) Bio-Impedance Measurement System for Analysis of Skin Diseases. International Journal of Application or Innovation in Engineering & Management, Vol. 3, Issue 2, February 2014.
7. Kumar S et al. (2012) Phase Angle Measurement in Healthy Human Subjects through Bio-Impedance Analysis. Iranian Journal of Basic Medical Sciences Vol. 15, No. 6, 1180-1184
8. Mialich MS, Faccioli JM Sicchieri et al. (2014) Analysis of Body Composition: A Critical Review of the Use of Bioelectrical Impedance Analysis. International Journal of Clinical Nutrition, Vol. 2, No.1, 1-10
9. AD5933 Application Notes at http://www.analog.com
10. Bertemes-Filho P, Felipe A and Vincence VC. (2013) High Accurate Howland Current Source: Output Constraints Analysis. Circuits and Systems, 4, 451-458
11. Kyle UG et al. (2001) Fat-Free and Fat Mass Percentiles in 5225 Healthy Subjects Aged 15 to 98 Years. Nutrition 17:534–541
12. Barbosa-Silva MCG et al. (2005) Bioelectrical impedance analysis: population reference values for phase angle by age and sex Am J ClinNutr2005; 82:49-52
13. Sun SS et al. (2003) Development of bioelectrical impedance analysis prediction equations for body composition with the use of a multicomponent model for use in epidemiologic surveys. Am J Clin Nutr, 77:331-40
14. Kyle UG. (2004) Bioelectrical impedance analysis. Part I: review of principles and methods. Clin Nutr, 23, 1226–1243
15. Chumlea WC et al. (2001) Total body water reference values and prediction equations for adults. Kidney International, Vol. 59, pp. 2250-2258
16. Beberashvili I et al. (2014) Bioimpedance phase angle predicts muscle function, quality of life and clinical outcome in maintenance hemodialysis patients. European Journal of Clinical Nutrition68, 683-689

Author: Fabrício Noveletto
Institute: State University of Santa Catarina (UDESC)
Street: Paulo Malschitzki, 200
City: Joinville
Country: Brazil
Email: fabricio.noveletto@udesc.br

Development of Portable Device to Measure Respiratory Activity Based on Impedance Pneumography

G.E. Cañadas, C.R. Dell'Aquila, and E. Laciar

Universidad Nacional de San Juan, Gabinete de Tecnología Médica, San Juan, Argentina

Abstract— **This paper describes the development of a portable device to register the respiratory activity based on the impedance pneumography method. The core of the system is the analog Front-End ADS1292R of Texas Instrument®, which incorporates all necessary stages for performing the measurement of the respiratory activity. The developed device includes a Bluetooth transmission for sending data to mobile phone or Personal Computer (PC). The system was tested with a multiparameter patient simulator and also with a healthy subject.**

Keywords— **Impedance Pneumography, Respiratory Signal, Embedded System, Analog Front-End.**

I. Introduction

The Respiratory Activity (RA) is a very important biomedical signal; however, it is not commonly monitored. Through the analysis of RA is possible to detect diverse respiratory pathologies, such as asthma, chronic obstructive pulmonary disease (COPD) and sleep apnea-hypopnea syndrome. Nowadays, different methodologies exist, direct and indirect, to RA measurement. One of the methods is based on the Impedance Pneumography (IP), which is an indirect technique for continuous and dynamic measurement of respiratory volume [1]. The IP is a specialized type of impedance plethysmography (IPG).

The IPG is a non-invasive method of determining changing tissue volumes in the body, based on the measurement of electric impedance at the body surface [2]. Under an alternating electrical excitation, the biological cells and tissues produce a complex bioelectrical impedance or electrical bioimpedance which depends on tissue composition and frequency of the applied AC signal. Therefore the frequency response of the electrical impedance of the biological tissues is highly influenced by their physiological and physiochemical status and varies from subject to subject [3].

Others physiological quantities estimated from electrical impedance measurement include blood flow, stroke volume, autonomic nervous system activity, muscle contraction, eye movement, endocrine activity and the activity of brain cells [4].

The IP were among the first applications of IPG. The objective of IP is to measure the tidal volume under resting conditions or during exercise. The principle is based on the measurement of the transthoracic impedance that increases during inspiration as a consequence of increasing alveolar air filling, and decreases during expiration [5].

The application of IP is very simple and also applicable for critically ill patients, since it allows continuous recording without requiring a breathing tube. However, the quantitative determination of the tidal volume is difficult and needs calibration by another Spiro metric method.

Previous work show that is possible diagnoses the COPD through the IP method and the results shows evidence that is possible to replace the spirometer for diagnosis and only use it a calibration method [6].

The aim of this work is to perform the development of a portable and low power device for RA measurement based on IP method. Unlike others similar works [6], the current system includes a (BT) transmission for send data to mobile phone or PC.

This paper is structured in the following way. Section II explains the operating principle of impedance pneumography method. Section III describes the implementation of the device hardware. Section IV shows the acquisition and validation of the respiratory signals. Finally, section V presents the results and conclusions.

II. Impedance Pneumography

The IP method consists of introducing an alternating current of high frequency (I_{AC}) in the tissues of the thorax through of electrodes. This current produces a voltage difference (V) between the electrodes which depends on the tissue impedance (Z). The measurement of the bioimpedance is based on Ohm's law, which is calculated as the ratio of the voltage difference between the electrodes and the current introduced into the tissue as $Z = V / I_{AC}$. The frequency of alternating current is in the range of 20-100 kHz [2]. This frequency is high enough to avoid stimulation of tissues, electrode polarization and excessively high skin impedance [7].

To apply this method there are two widely used electrode configurations; bipolar and tetrapolar. In this paper, we used the bipolar configuration (Fig. 1), because it requires only two electrodes for measuring impedance, and hence the

F. Simini and P. Bertemes-Filho (eds.), *II Latin American Conference on Bioimpedance*,
IFMBE Proceedings 54,
DOI: 10.1007/978-981-287-928-8_16

current signal injection and voltage measurement are conducted with the same electrodes. The two-electrode method, therefore, suffers some error from the contact impedance problem and the measured data contains the voltage drop due to the contact impedance [3].

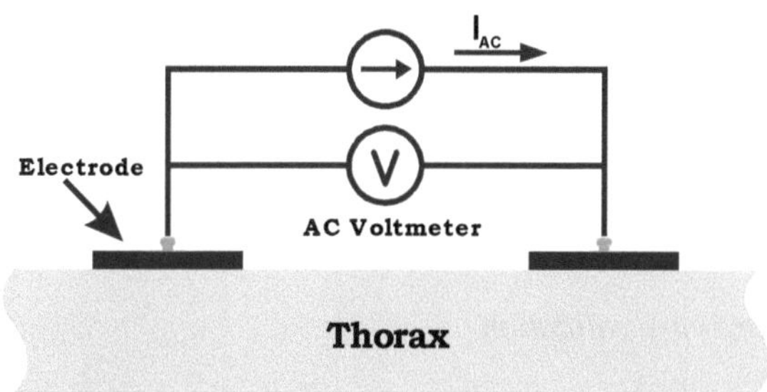

Fig. 1 Bipolar electrode configuration (Two-electrode measurement).

When measuring respiration, the thorax presents electrical impedance to the electrode that consists of two impedance components (Fig. 2). One of the components is a relatively constant value, which is commonly known as the baseline impedance or R_B and his average value is around 500Ω [7-8]. The other component is a varying value, known as respirative impedance or ΔR. This component of impedance generates a varying voltage component when current is injected. This varying voltage component is the parameter of interest because this component can then be used to determine the person's breathing rate.

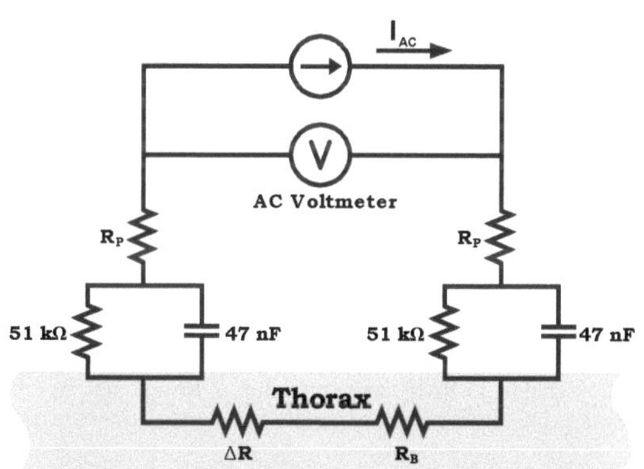

Fig. 2 Electrical model of the respiration circuit.

The electrical model of the respiration circuit of the Fig. 2 includes the defibrillator protection resistances (R_P) in each wire of the ECG cable and the electrode impedance. The electrode impedance is modeled as a 51-kΩ resistor in parallel with a 47-nF capacitor as specified in the ANSI/AAMI EC13:2002 standard. The total baseline impedance consists of the sum of R_P, R_B, and electrode impedance [9].

III. HARDWARE IMPLEMENTATION

The hardware was implemented based on integrated circuit (IC) ADS1292R from Texas Instrument®. It is an analog Front-End that incorporates: a *Modulator* (MOD), a *Programmable Gain Amplifier* (PGA), a *Demodulator* (DEMOD), a 24 bits *Analog to Digital converter* (ADC) and a *Serial Peripheral Interface* (SPI) communication to send digitalized signals to the microcontroller (MCU), in order to measure the IP. The IC has two ECG channel but this function hasn't been used in this work. Fig. 3 shows the block diagram of the system.

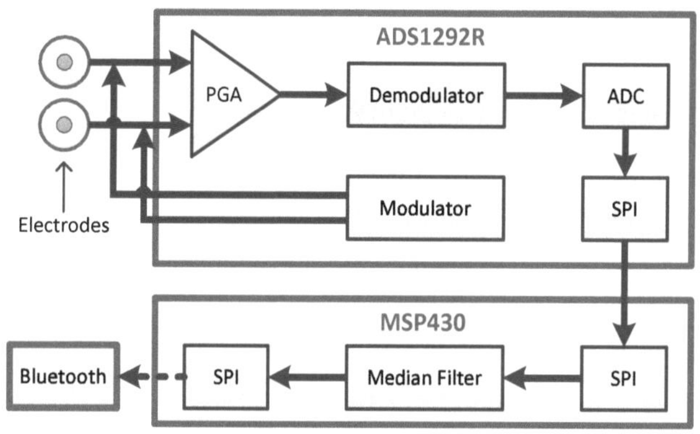

Fig. 3 Block diagram of hardware implementation

The Modulator generates a square wave of 32 kHz that is used as a carrier signals of current (I_{AC}) and it is connected to the electrodes and then it is injected to the thorax. The current is 100μA in accordance with the ANSI/AAMI ES1-1993 standard. This signal is Amplitude-Modulated (AM) by the low-frequency signal (ΔR) generated as a results of the RA.

The electrodes are connected to the inputs of PGA, too. It amplifies the AM voltage signal from the thorax. The gain of the amplifier can be configured between 1 to 12. A lower gain let saving power.

The function of the DEMOD block is demodulating the AM signal from PGA and the type of the demodulation process is synchronous. This process consists in multiply AM signal by demodulating signal. It is simply a square wave signal at the same frequency as the carrier signal. The phase of this must be adjusted in accordance with AM signal delays. The ADS1292R allows us choice between various fixed values of phases. Set correct phase is important; otherwise the carrier signal wouldn't be eliminated in the demodulation process. Fig. 4 shows the difference between correct and incorrect phase selection.

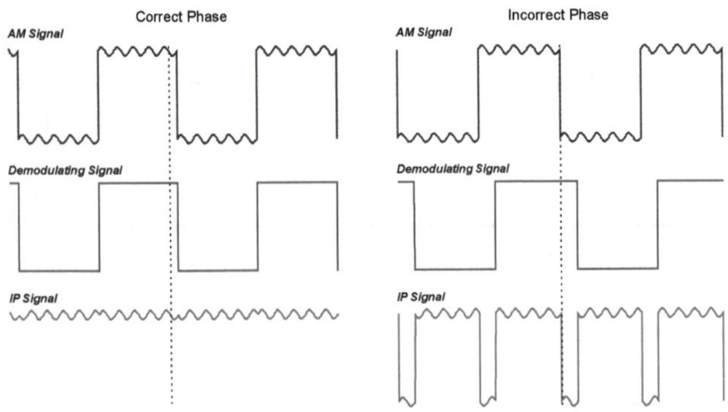

Fig. 4 Effect of the incorrect demodulation phase selection

Other problem in demodulating process is that square waveform carrier is modified by electrical path then it is different to demodulating signal.

If the sinus as carrier and demodulating signal would be used, the problem with phase would be the same but the carrier wouldn't be modified by electrical path. The DEMOD block sends the demodulated signal to ADC. This is proportional to IP signal. The complete process of modulation and demodulation is showed in Fig. 5.

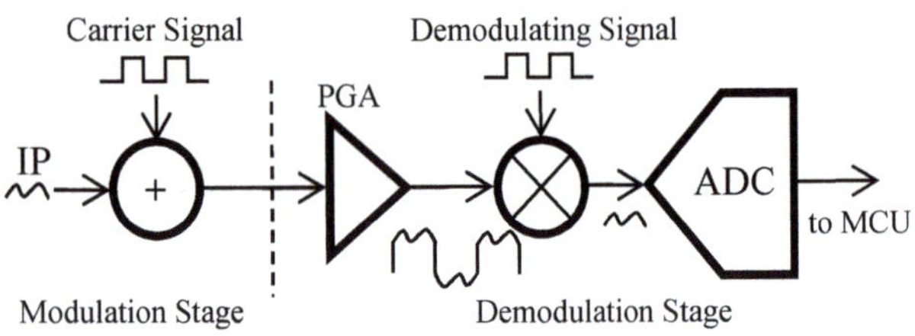

Fig. 5 Modulation and Demodulation process

The ADC digitalizes the IP signal and sends data, through SPI, to MCU. The sample rate (SR) is greater (250Hz) than respiratory signal's spectrum [10] because the noise can have high frequency components. However the signal can be decimated after filtering for saving power in BT transmission.

The ADC is 24bits. The use of high resolution ADC allows to have low gain amplifiers and therefore low power consumption. It can digitally represent low voltage signals. The MCU received IP signal and filter it with a Median Filter (MF). Then the data is sent, through BT, to a mobile device. The device uses the MSP430 MCU. It is low power and suitable for wearable applications. The BT module used is a generic controller based on CC2560 IC from TI.

The MF is a no-linear filter used to remove impulse noise and consists in perform the statistical median between current sample and its neighborhood. The number of samples that are include in it form a temporal window. The filter is characterized by its windows and in this work is 1 second.

This value is because the normal respiration is between 12 and 20 respirations per minute (RPM) [11] then the maximum period of the respiratory signal and IP signal would be greater than filter window (3s).The MF is used to remove noise from harmonics carrier and other biological signals like electrocardiographic (ECG) and electromyographic (EMG) signals [12].

IV. Signal acquisition and validation

A. Hardware validation

To validate the performance of the hardware model a multiparameter patient simulator PS420 from Fluke Biomedical® was used, calibrated at the time of the test. It allows simulate the IP signal. The respiratory rate (RR), R_B and ΔR can be set in the simulator. The test performed with fixed R_B in a typical value (500Ω) as it was mentioned. The ΔR was set the minimum possible value (0.2Ω), because this represents the worst condition for the device. The Fig. 6 shows IP signal acquired with 15 (RPM) and the filtered signal with MF. The phase was set in 12°, a recommend value by the manufacturer of CI.

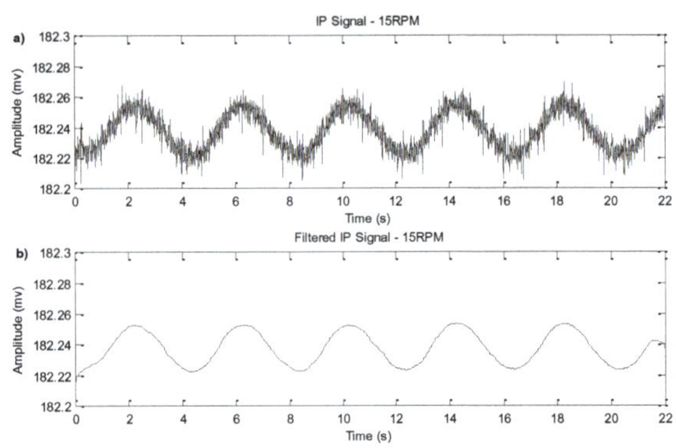

Fig. 6 IP signal of 15 RMP from simulator. a) Raw signal. b) Filtered signal

B. Real signal acquisition

In order to evaluate the overall system performance, a measurement on a healthy subject was also performed. The device acquired the signal and send digitalized data to PC, via Bluetooth. Fig. 7 shows acquired raw and filtered signal. In the first case can be observed that the signal has the QRS complex from heart activity. The filter removes this and other spurious. The electrodes placements are right and left mid-axillary line based on EASI lead system that is used with holters [13].

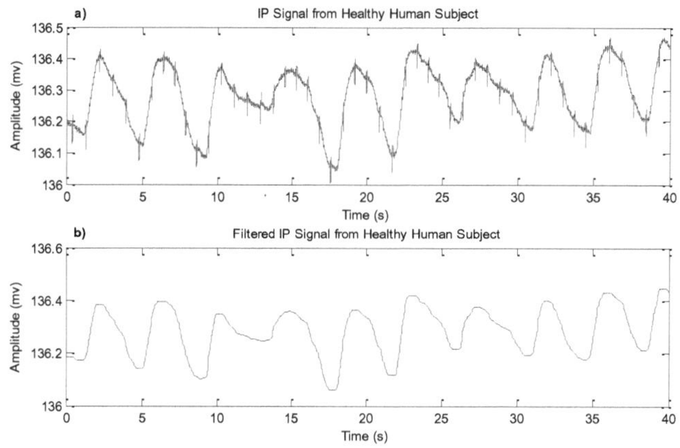

Fig. 7 : IP signal from healthy subject. a) Raw signal. b) Filtered signal.

The Fig. 7 shows the amplitude of signals in mV related with tildal volume but the system hasn´t been calibrated with for example Spirometer.

V. Discussion and conclusions

A portable device for monitoring RA has been developed in this work. It is based on IP, which is an indirect method for measured the RA. The system was implemented with the IC ADS1292R which uses a square waveform as carrier signal, unlike the traditional methods that use a sinus waveform. The noise generated by harmonics of the square signal and others biological signal like ECG and EMG signals, are removed with MF.

The device has a low power MCU for read digitalized IP signal from ADC by SPI interface. The MF and controller of the Bluetooth module is implemented in the MCU, too.

The signal is transmitted to a mobile phone or PC for storage or analysis purpose.

A multiparameter patient simulator was used for performance validation. It can generate signals with different respirations rate. Fig. 6 shows the signal acquired with typical values and it can be observed that signal period is the value configured in the simulator. The noise reduction after filter can be appreciated too.

On the other hand, the performance of the system was tested with a healthy subject, Fig. 7 shows the results. The noise generated by heart activity is eliminated with the filter.

In conclusion, a low cost and power device was implemented based on a simple and traditional method for evaluating the respiratory activity.

In futures works four-electrodes method will be tested for improve signal to noise ratio. The calibration of the system will be performed in futures works too, using a Spirometer.

Acknowledgment

All authors are supported by Consejo Nacional de Investigaciones Científicas y Técnicas (CONICET). The device developed in this work was supported by grants of Agencia Nacional de Promoción Científica y Tecnológica and Universidad Nacional de San Juan (ANPCyT – PICT-O-UNSJ 2009 N° 0027). All institutions are from Argentina.

Conflict of Interest

The authors declare that they have no conflict of interest.

References

1. Pacella A. F. (1966) Impedance pneumography—a survey of instrumentation techniques, Med. & Biol. Eng., 4, 1–15.
2. Malmivuo J., Plonsey R. (1995) Bioelectromagnetism - Principles and Applications of Bioelectric and Biomagnetic Fields, Oxford University Press, New York.
3. Bera T. K. (2014) Bioelectrical Impedance Methods for Noninvasive Health Monitoring: A Review, J Med. Eng., vol. 2014, Article ID 381251, 28 pages,. DOI 10.1155/2014/381251.
4. Geddes L. A., Baker L. E. (1975) Principles of applied biomedical instrumentation, 2nd Edition, Wiley-Interscience, New York.
5. Nopp P, Rapp E., et al. (1993) Dielectric properties of lung tissue as a function of air content, Phys. Med. Biol. Vol. 38, pp 699–716.
6. Sim M. H., et al. (2013) Development and Evaluation of an Improved Technique for Pulmonary Function Testing Using Electrical Impedance Pneumography Intended for the Diagnosis of Chronic Obstructive Pulmonary Disease Patients, Sensors, 13(11), pp. 15846-15860
7. Grenvik A., Ballou S., McGinly E., et al. (1972). Impedance pneumography: Comparison between chest impedance changes and respiratory volumes in 11 healthy volunteers, Chest, vol. 62, pp 439-443.
8. Severinghaus J. W. (1971) Electrical measurement of pulmonary oedema with a focusing conductivity bridge, J Physiol., vol. 215, pp 53-55.
9. Gupta A. K. (2011) Respiration Rate Measurement Based on Impedance Pneumography, Texas Instruments application report SBAA181.
10. Redmond S. J.,Heneghan C. (2006) Cardiorespiratory-Based Sleep Staging in Subjects With Obstructive Sleep Apnea, IEEE Trans. Biomed. Eng., vol. 53, pp 485-496.
11. Guyton, A., & Hall, J. (2010). Textbook of medical physiology (12th ed.), Saunders, Philadelphia.
12. Chazal P., Penzel T., Heneghan C. (2004) Automated detection of obstructive sleep apnea at different time scales using the electrocardiogram, Physiol. Meas., vol. 25, pp 967-983.
13. Jahrsdoerfer M., Giuliano K., Stephens D. (2005) Clinical Usefulness of the EASI 12-Lead Continuous Electrocardiographic Monitoring System, Clinical Journal of the American Association of Critical-Care Nurses. 5 vol. 25 pp. 28-37.

Author: Gabriel E. Cañadas
Institute: Universidad Nacional de San Juan - GATEME
Street: Av. Libertador 1109 (oeste) – CPA: J5400ARL
City: San Juan
Country: Argentina
Email: gcanadas@unsj.edu.ar

Signal Processing Architecture for Electrical Tomography Impedance

Raul Gonzalez Lima, André Luis dos Santos, Erick Dario León Bueno de Camargo,
Fernando Silva de Moura, and Talles Batista Rattis Santos

Laboratório de Engenharia Ambiental e Biomédica, Department of Mechanical Engineering,
Polytechnic School of University of São Paulo, São Paulo, Brazil

Abstract— **This paper presents a description of an architecture of an EIT hardware for research. Both, hardware and software for capturing and processing of the EIT signals are addressed. The system is divided in modules with defined requirements and connections, therefore, different implementations are possible. Details of an implementation conceived to validate the architecture with respect to processing speed is also described.**

Keywords— **EIT, microcontroller, signal processing, communication**

I. Introduction

Electrical Impedance Tomography (EIT) is a technique for obtaining internal images of a domain by measuring and exciting electrical signals within its boundary. Images are obtained by estimating the domain impedance distribution. Besides accuracy, precision and high speed, an EIT measuring system has additional requirements. There are safety limits related to the electrical potentials imposed to the human body. Proper insulation, contact impedance and positioning of the electrodes are also relevant matters in a tomograph to be used on *in vivo* data acquisition conditions.

This paper presents an architectural description of an EIT instrumentation design. This architecture was conceived after the analysis systems reported in the literature, remarkably the systems described in [1], [2], [3], [4] and [5], and the EIT system used at the Laboratório de Engenharia Ambiental e Biomédica of the Polytechnic School of University of São Paulo. The main goals are:

- Allowing the use of different demodulation methods or raw signal analysis;
- Compatibility with different signal conditioning circuits implementations;
- Accomplishing speed and accuracy required for 50 images/second with 32 electrodes.
- Defining a deterministic communication process for data transfer.

II. Proposed architecture

The proposed architecture, presented in Fig. 1, defines a set of semi-independent modules that, working together provides EIT requirements. There are Measuring Channels which capture and process the electrodes signals, an Excitation System responsible for imposing current excitation patterns and a Communication Network, which communicates measurements to a Supervisor module that will act like an interface for an Impedance Estimation Hardware and Software.

For each electrode there is a Measuring Channel also connected to the Communication Network. This, theoretically, allows a large number of electrodes. However, practical application will be limited by the Communication Network speed and processors capacity to handle the information.

In the sequel each module requirements and connections are described. Some details of built experimental circuit implemented for validation is also presented.

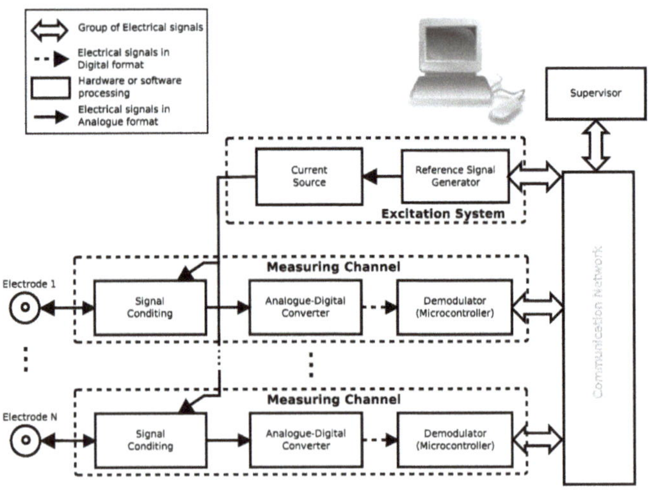

Fig. 1 Modular architecture proposed

F. Simini and P. Bertemes-Filho (eds.), *II Latin American Conference on Bioimpedance*,
IFMBE Proceedings 54,
DOI: 10.1007/978-981-287-928-8_17

A. Measuring Channel

Each Measuring Channel is composed by Signal Conditioner, Analogue-Digital Converter and Demodulator. This set of modules deliver three scalar information, namely, amplitude, phase and quality, for every acquisition cycle. It was defined an acquisition cycle of 625 μsec to attend the requirement of 50 images/second.

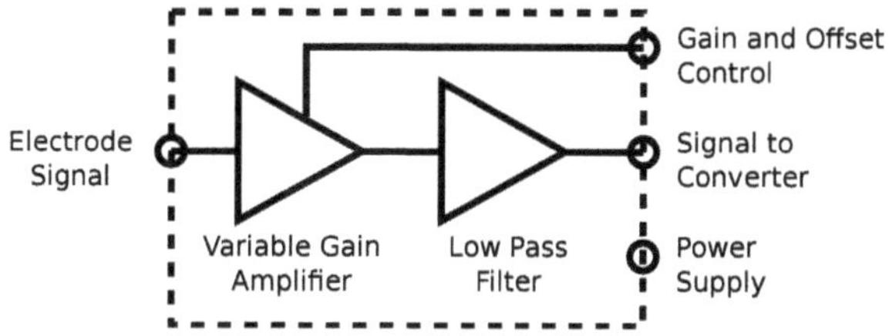

Fig. 2 Signal Conditioner module

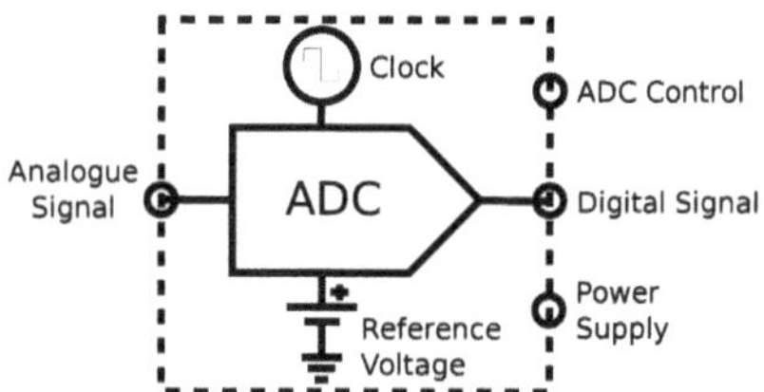

Fig. 3 Analogue-Digital Converter module

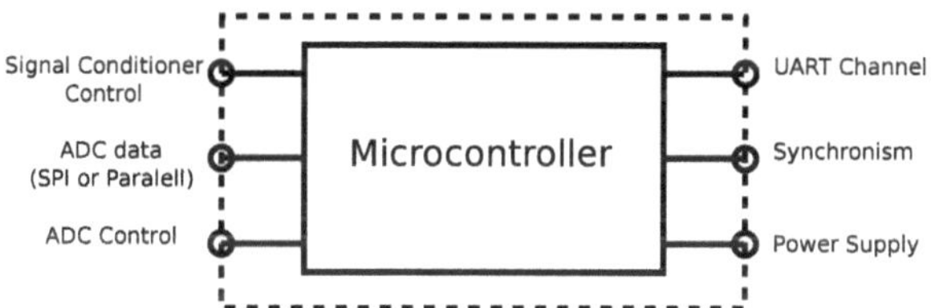

Fig. 4 Demodulator module

The Signal Conditioner must implement an anti-aliasing filter and add DC offset to the electrode signal as shown in Fig. 2. The electrode is directly connected to this module input, therefore, a high input impedance is necessary. Filtering should be done with minimal phase/amplitude distortion and maximal temperature stability. An automatic gain control (AGC) managed by the Demodulator improves the system measuring accuracy as intensity of signals are proportional to the electrode distance from excitation point.

The Analogue to Digital Converter module is basically a commercial IC and its accessories circuits (clocks, references, etc.). Its connections are shown in Fig. 3.

The Demodulator module is a microcontroller with necessary hardware to read and control ADC of the Analogue-Digital Converter module, control Signal Conditioner AGC and handle the communication with the Supervisor module, as shown in Fig. 4. It also must have the processing capacity to execute, each acquisition cycle, the following tasks:

1. *ADC Reading* - Each ADC reading must be stored.

2. *Phase Synchronism* - Excitation System generates a clock which must be used for synchronize ADC reading and Demodulation.

3. *Demodulation* - ADC data must be demodulated in order to find its amplitude and phase. An additional value will indicate signal to noise ratio.

4. *Signal Conditioner Control* - Initial amplification must be controlled in order to keep analogue signal near full range of the ADC.

5. *Data Transfer* - Transfer to Supervisor runs asynchronously. This communication demands processor speed and memory.

6. *Calibration* - Some additional calibration tasks will require program memory.

B. Excitation System

The Excitation System is composed by a Current or a Voltage Source and a Reference Signal Generator. The Reference Signal Generator is responsible for the generation of the signals for synchronism of the system. These signals controls the excitation pattern, acquisition period and phase demodulation. Its connections are shown in Fig. 5.

The Current Source module ensures an invariant current amplitude flowing through a pair of electrodes. Digitally controlled analogue switches allow the selection of this pair, defining a particular excitation pattern.

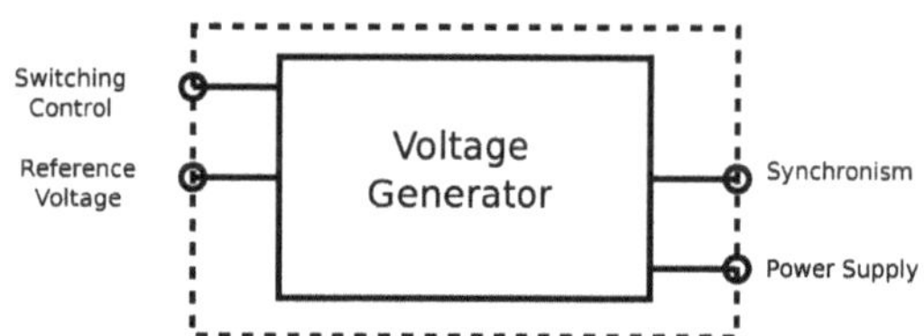

Fig. 5 Reference Signal Generator module

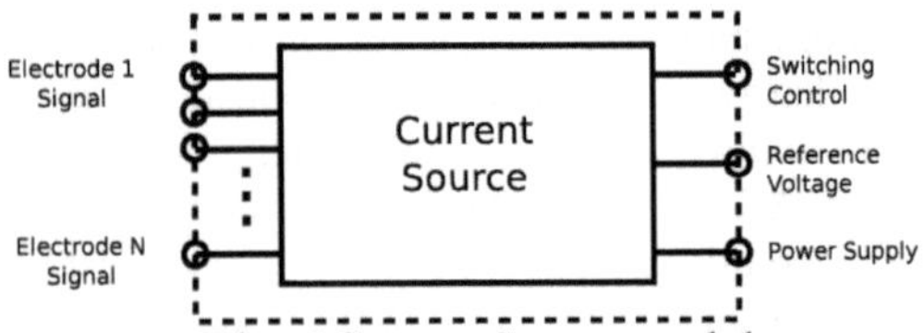

Fig. 6 Current Source module

C. *Supervisor*

The Supervisor module is a software running in a Personal Computer (PC). This software stores demodulated digital data and stores calibration data, for instance. Standard USB or Ethernet ports and storage space are the only requirements in the computer hardware as regular processors are fast enough to handle the required data processing.

D. *Communication Network*

Since the Supervisor module requires USB or Ethernet connection and this protocol introduces an unnecessary overloading of processing, an additional microcontroller is used to reduce complexity and costs. This microcontroller is connected by USB or Ethernet to the computer and handles a network of the Measuring Channels.

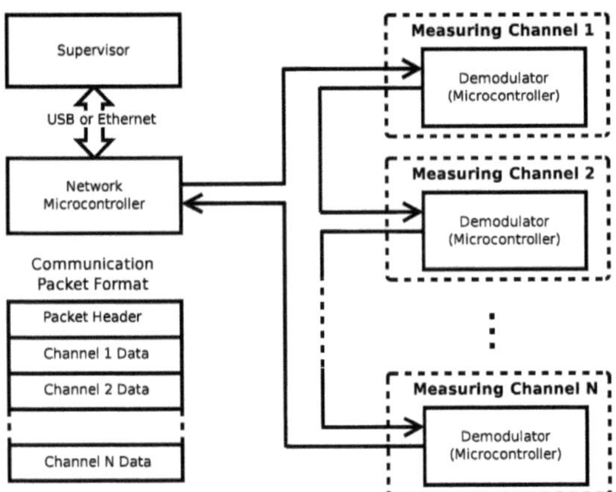

Fig. 7 Network structure and packet format

A ring topology was chosen instead of a bus topology to avoid medium access control or master/slave processing. A star topology would require more communication channels than normally available on commercial microcontrollers. The Fig. 7 shows the connections and the packet format. The Network Microcontroller creates the packet with the proper header to inform channels about information required or operation to be performed. As the packet is replied by the Measuring Channels, they fill the Channel Data area.

III. RESULTS AND DISCUSSION

An implementation of the proposed architecture was created. This implementation uses an ADC of 16 bits and 2.5M samples/s (AD9260, Analog Devices). The Fig. 8 shows the Analogue-Digital Converter module implementation. A 20MHz crystal and the IC AN780 (Analog Devices) were added for, respectively, clock and voltage reference.

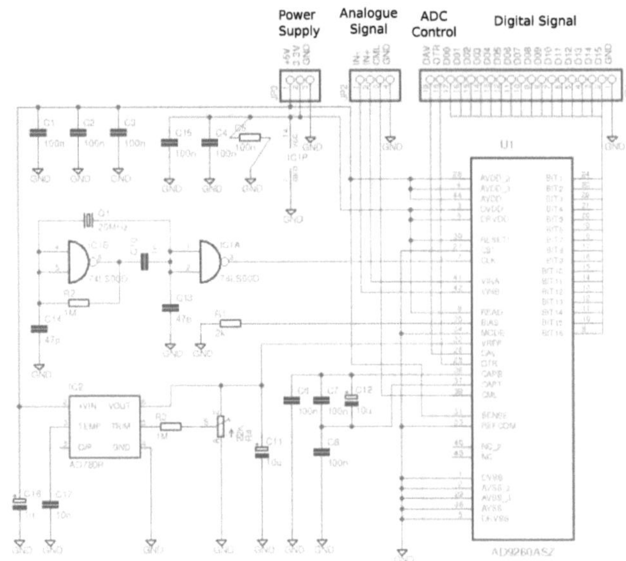

Fig. 8 Analogue-Digital Converter Implementation

A 32 bits ARM microcontroller (STM32F407, ST Microelectronics) was used as Demodulator and Network Hub. USB was the choice to communicate with the PC and 10.5 Mbit/s UART as Measuring Channels network. The traffic of information required was achieved as shown the Fig. 9. It was accomplished because the demodulation process performed in the Measuring Channel reduces 1000 times the volume of data.

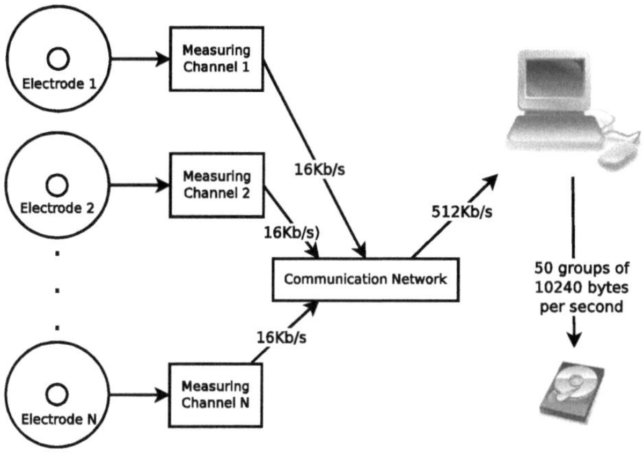

Fig. 9 Information flow

The Signal Conditioner module was not implemented with proper components and proper shielding. The resulting measurements, as shown in Fig. 10, were not within the required accuracy. The main goal of this experimental implementation is to show that the speed of signal processing is achievable with the proposed architecture.

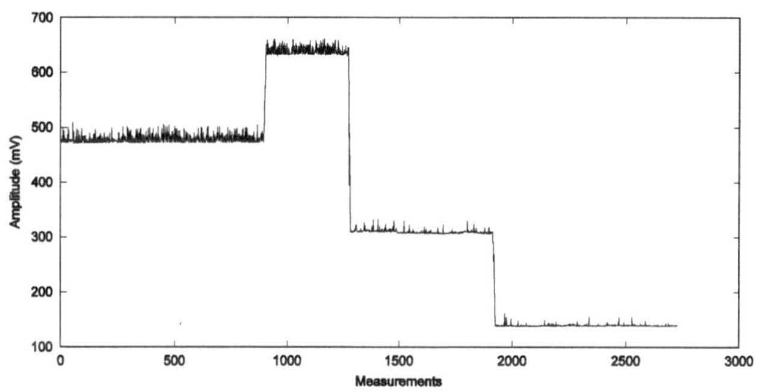

Fig. 10 Results with four different amplitude signals

IV. Conclusion

The proposed architecture was implemented, experimentally tested and the required processing speed was achieved. A more careful implementation of the Signal Conditioner module is necessary to obtain the required accuracy.

V. References

[1] G. J. Saulnier, N. Liu, C. Tamma, H. Xia, T.-J. Kao, J. C. Newell, and D. Isaacson, "An electrical impedance spectroscopy system for breast cancer detection," Proceedings of the 29th Annual Inter-national Conference of the IEEE EMBS, 23 aug 2007.

[2] N. Liu and R. P. Institute, ACT4: A High-precision, Multi-frequency Electrical Impedance Tomograph. Rensselaer Polytechnic Institute, 2007.

[3] T. I. Oh, H. Wi, D. Y. Kim, P. J. Yoo, and E. J. Woo, "A fully parallel multi-frequency eit system with flexible electrode configuration: Khu mark2," IOP Publishing - Physiological Measurement, vol. 32, 7 jun 2011.

[4] A. P. Moumbe, "Central unit development of a simultaneous aquisition system for eeg and eit data," Master's thesis, University of Montreal, dec 2011.

[5] A. P. M. R Guardo, J Jehanne-Lacasse and H. Gagnon, "System front-end design for concurrent acquisition of electroencephalograms and eit data," International Conference on Electrical Bioimpedance, 2014

In vivo Electrical-Impedance Spectroscopy (EIS) Readings in the Human Rectum

Mulett-Vasquez Edelberto[1], Gonzalez-Correa Carlos-Augusto[2], Miranda-Mercado David-Alejandro[3], Osorio-Chica Mauricio[1], and Dussan-Lubert Carmen[4]

[1] Universidad de Caldas, Department of Surgery, Manizales, Colombia
[2] Universidad de Caldas, Director of the Research Group on Electrical Impedance, Manizales, Colombia
[3] Universidad Industrial de Santander, Department of Physics, CIMBIOS, Bucaramanga, Colombia
[4] Universidad de Caldas, Department of Mathematics, Manizales, Colombia

Abstract- **Worldwide, colorectal cancer (CRC) is one of the leading causes of morbidity and mortality, ranking third among the most common cancers. Although screening programs are useful in reducing its impact, the screening tools available at present are far from being ideal. This justifies the search for new screening tools, and the authors think that Electrical Impedance Spectroscopy (EIS) could be a good candidate. It is a non- or minimally-invasive technique, safe, low cost and easy to use. At present, it is being introduced in cervical cancer screening programs in the UK, and has been tested in other types of cancer. In this preliminary study with a small sample of 17 human patients undergoing diagnostic total colonoscopy, we report the feasibility of using EIS for taking *in vivo* rectal readings. Four quadrant measurements were taken on each volunteer, a procedure that, in total, takes no more than 2 minutes, with minimal additional discomfort for the volunteers. After colonoscopy, 10 volunteers were considered as normal, 5 had polyposis, 1 had diverticulosis and 1 had disuse colitis. None was diagnosed as having cancer anywhere in the colon or rectum. A Multiple Range Test with Tukey HSD (Honest Significant Difference) shows a statistical significant difference between the readings obtained from the volunteer with disuse colitis when compared to the readings obtained from patients of the other three groups. *In vivo* EIS readings in humans can be obtained very easily and this opens the possibility of using it for further studies in the search of its possible application as an effective screening tool.**

Keywords-- **Colorectal Cancer, Electrical Impedance Spectroscopy, Screening Tools.**

I. INTRODUCTION

With 9.7% (1.361.000) of all new cancers diagnosed worldwide in 2012, colorectal cancer (CRC) is the third most common cancer [1]. As with many other cancers, in the search for reducing CRC associated morbidity and mortality, screening programs have been implemented in different parts of the world [2]. Although there are different tests available for early diagnosis of CRC, screening programs usually use one or both of the two following: detection of fecal occult blood (FOB) and endoscopy, either by flexible sigmoidoscopy or by colonoscopy [2] [3]. The former has as major disadvantage, its low sensitivity for both CRC and advanced adenomas. While the later is invasive, expensive, time-consuming, associated with possible pain, requires rigorous peroral bowel cleansing and good trained specialized personnel [2], making it difficult to implement worldwide. Ideally, a screening test should be able to detect not only CRC at an early stage, but rather identify persons with cancer-prone colorectum, so that prevention and closer follow up programs can be implemented [4] [5]. In the literature, there is published evidence that increased transepithelial permeability, leading to a decrease of its electrical impedance, is associated with the development of cancer [4] [6] [7] [8]. However, it seems that, regrettably, this line of research has not been explored further after this pioneering work. EIS is an inexpensive, easy to use, non- or, at most, minimally-invasive technique that is being commercialized in the UK and other parts of the world to improve the diagnosis of cervical cancer as an adjunct to colposcopy [8]. It has also been used in research for early detection of cancer in other organs as oesophagus [10], bladder [11] and skin [12]. In this preliminary report, we show initial results of an ongoing project on the application of EIS for early detection of CRC. Through readings taken from the distal rectum, with a procedure that takes no more than 2 minutes, where the patient doesn't need preparation/sedation. It can be considered as a completely safe and minimally invasive procedure, with no need for a specialist being involved and of very low cost. The main aim of this paper is to report these findings, showing the feasibility of this approach.

II. MATERIALS AND METHODS

Volunteers were recruited among patients sent for diagnostic colonoscopy to "Union de Cirujanos SAS" at Clinica La Presentación in Manizales, Caldas, Colombia, South-America, between December 2014 and February 2015. Readings were collected from 17 volunteers attended by the endoscopist surgeon involved in the project (MOCh), after signing the informed consent approved by the Ethical Committee from the University of Caldas. Gender, number of subjects, and ages of the sample are shown in table 1.

Table 1. Gender and ages of the volunteers

Gender	n	Median age	Minimum age	Maximum age
Female	12	48,5	22	67
Male	5	64,0	24	67
Total	17	49,0	22	67

F. Simini and P. Bertemes-Filho (eds.), II Latin American Conference on Bioimpedance,
IFMBE Proceedings 54,
DOI: 10.1007/978-981-287-928-8_18

Using the same equipment as Brown *et al* [13] [14] and Gonzalez *et al* [15], and with the same type of pencil-probe used on the cervix, an electrical current of 40 µA peak-to-peak at 7 different frequencies was passed between two adjacent electrodes. The probe had a diameter of 5.5 mm with four gold electrodes (1 mm diameter each) equally spaced on a circle of radius 1.65 mm and mounted flush with the face of the probe. The 7 frequencies used went from 9.6 kHz to 614.4 kHz, each frequency being the double of the previous one. The real part of the resulting potential was measured between the two remaining electrodes and the transfer impedance was calculated as the ratio of the measured potential to the amplitude of the imposed current. Electrical readings were taken prior to colonoscopy by the colorectal surgeon participating in the project (EMV). With the patient in fetal position and after lubrication of the anus, a disposable anoscope was gently introduced. The surgeon proceeded then to take measurements at positions 00:00 (posterior, i.e. middle line immediately below the coccyx), 03:00 (right lateral), 06:00 (anterior), and 09:00 (left lateral) (Figure 1.). Subsequently, the colonoscopy was carried out but no biopsies were taken from the rectum, as that would be inacceptable from the clinical and ethical point of view.

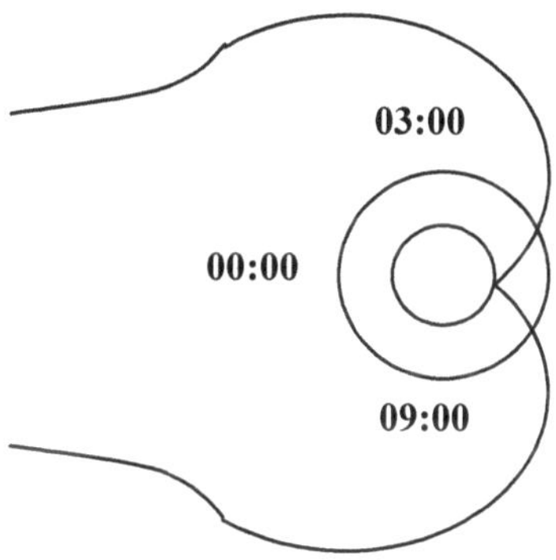

Figure 1. Positions where readings were taken. See text for more details.

The Student's t-Test was used to assess the statistical significance of the differences between the resistances at the different frequencies, on the different points. To assess the differences between data of the four groups of patients in the sample (colitis, diverticulosis, polyps and normal), we used the Multiple Range Test with Tukey HSD (honest significant difference).

III. RESULTS

The first answer that we wanted to resolve was if there was any statistical significant difference in the readings obtained in the four different sites described above, as it is shown in Figure 1. The averages of all readings at each site and at all frequencies are practically the same (Figure 2) and there is no statistically significant difference between points of measurement at each frequency (Table 2).

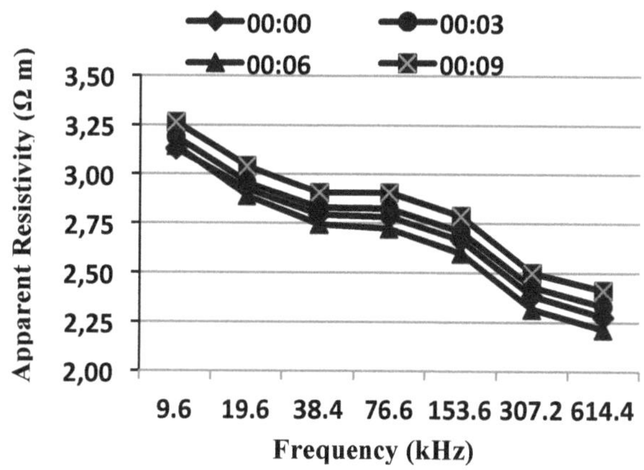

Figure 2. Electrical Impedance Spectra of the mean resistivity at each frequency for the four different positions where readings were taken (n = 17).

Table 2. P values for the differences between the means of the readings obtained on the four point at seven frequencies.

Frequency (kHz)	P-Value
9.6	0,90
19.6	0,86
38.4	0,83
76.6	0,77
153.6	0,72
307.2	0,64
614.4	0,60

After colonoscopy, of these initial 17 volunteers, 10 were considered as normal (no apparent lesion), 5 had polyposis, 1 had diverticulosis and 1 had disuse colitis. The differences between the averages of the readings from patients with polyposis, diverticulosis and normal are statistically no-significant (Figure 3). Table 3, shows the mean of the readings for each group of volunteers, where those pertaining to diverticulosis, colitis and normal are homogeneous, i.e. they can be considered as equal. The Multiple Range Test with Tukey HSD shows that the difference is significant when Colitis-Diverticulosis, Colitis-Normal and Colitis-Polyposis are compared (Table 4).

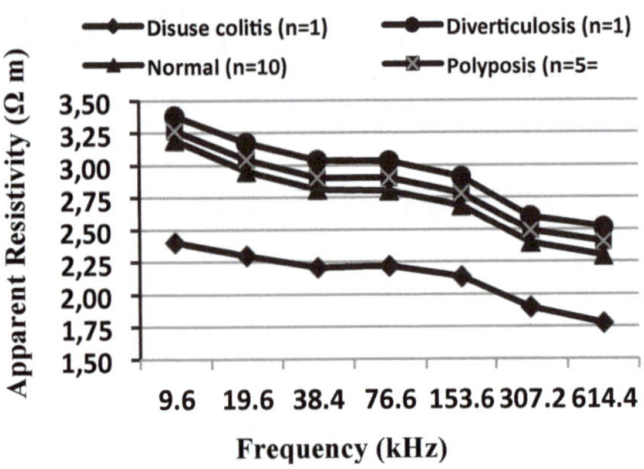

Figure 3. Electrical Impedance Spectra of the mean resistivity for the four groups of volunteers.

Table 3. Mean of the readings obtained in each group.

Group	Total readings	Mean
Colitis	28	3,88
Diverticulosis	28	4,95
Normal	280	5,06
Polyposis	140	5,27

Table 4. Comparison of the means between the four groups of patients.

Contrast	Difference	+/- Limits
Colitis - Diverticulosis	*-1,06	0,69
Colitis - Normal	*-1,17	0,51
Colitis - Polyposis	*-1,39	0,53
Diverticulosis - Normal	-0,11	0,51
Diverticulosis - Poliposis	-0,33	0,53
Normal - Poliposis	-0,22	0,27

* denotes a statistically significant difference.

IV. Discussion

The number of subjects of the study is very small, but, as its main aim was to see the viability of obtaining the readings, we think that it is enough for that purpose. Obtaining EIS readings from patients undergoing diagnostic total colonoscopy was carried out very easily, with practically very little additional discomfort for the subjects. In all cases, it took, in total, less than two minutes to have the readings in the four selected positions. Although there seems not to be statistically significant differences among the readings obtained in them, we still consider that, given the easiness of the procedure, it is worth having all four measurements and averaging them to obtain a single value for each patient. In his PhD thesis, Gonzalez-Correa [16] reported median values of apparent resistivity in columnar epithelia from the stomach and from esophageal areas with intestinal metaplastic mucosa (Barret´s esophagus) at the five lower frequencies used in this study (9.6, 19.6, 38.4, 76.6 and 153.6 kHz) of 4.2, 4.0, 3.7, 3.4, 3.1 Ω m, respectively. The medians of the readings obtained for the rectal mucosa from all subjects in this study gave an averaged apparent resistivity of 3.2, 2.9, 2.8, 2.8, 2.7, 2.4 and 2.2, Ω m at the same frequencies. Our values are lower (almost by a third), but this is something that could probably be explained by the presence of a thicker layer of mucus in the colon. Values for this variable have been reported to go up to 800 μm in the colon of some experimental animals [17]. Another factor that could contribute to these differences is the size of the probes used in both studies, as the one used in the esophagus was thinner (diameter of 3.2 mm with electrodes with diameter of 0.8 mm, equally spaced on a circle of radius 1.35, compared to ours with a diameter of 5.5 mm and four gold electrodes with diameter of 1 mm each, the same used by Brown *et al* [13] in the human cervix. Based on Geselowitz´s theory [18], it is known that the size and the separation of the electrodes influence the penetration depth of the applied current [16,19].

V. Conclusions

With this initial small study, we now know that it is feasible to obtain EIS in vivo readings from the rectum of human patients undergoing colonoscopy. As, if this technique were to be used on its own, there would be no need, neither for previous preparation of the patients, not for sedation, and the examination is minimally invasive, safe, easy to perform and consumes very little time. It makes this technique a very good candidate as a possible screening tool. We will explore this possibility in the future, and will also consider the use of different probe sizes, as they can give information about structures at different depths of the rectal wall.

Aknowledgments

The authors gratefully thank to Unión de Cirujanos SAS, especially to Mr. Lázaro-Antonio Arango-Molano for authorizing to carry out this initial study in their facilities.

Conflict of Interest

The authors declare no conflict of interest.

Compliance of Ethical Requirements

All subjects gave their informed written consent in accordance with the guidelines of the Ethical Committee of the University of Caldas.

References

1. Ferlay J, Soerjomataram I, Ervik M, et al. GLOBOCAN 2012 v1.0, Cancer Incidence and Mortality Worldwide: IARC CancerBase No. 11 [Internet]. Lyon, France: International Agency for Research on Cancer; 2013. Available from: http://globocan.iarc.fr, accessed on 20/03/2015.
2. Garborg K, Holme Ø, Løberg M, et al. (2013) Current status of screening for colorectal cancer. Ann Oncol 24:1963-1972.
3. Rozen P Winawer S Waye J (2002) Prospects for the worldwide control of colorectal cancer through screening. Gastrointest Endosc 555:755-759.
4. Davies R, Joseph R, Asbun H, et al. (1989) Detection of the cancer-prone colon, using transepithelial impedance analysis. Arch Surg 124:480-484.
5. Rozen P. (2004) Cancer of the Gastrointestinal tract: early detection or early prevention?. Eur J Cancer Prev 13:71-75.
6. Davies R, Joseph R, Kaplan D, et al. (1987) Epithelial impedance analysis in experimentally induced colon cancer. Biophys J 52:783-790.
7. Soler A, Miller R, Laughlin K, et al. (1999) Increased tight junctional permeability is associated with the development of colon cancer. Carcinogenesis 20:1425-1431.
8. Davies R, Sandle G, Thompson S. (1991) Inhibition of the Na+,K(+)-ATPase pump during induction of experimental colon cancer. Cancer Biochem Biophys 12:81-94.
9. Tidy J, Brown B, Healey T, et al. (2013) Accuracy of detection of high-grade cervical intraepithelial neoplasia using electrical impedance spectroscopy with colposcopy. BJOG 120:400-410.
10. Gonzalez-Correa CA, Brown BH, Smallwood R, et al (2003) Low frequency electrical bioimpedance for the detection of inflammation and dysplasia in Barrett's oesophagus. Physiol Meas 24:291-296.
11. Keshtkar A, Mesbahi A, Mehnati P, et al. (2008) Surface fluids effects on the bladder tissue characterisation using electrical impedance spectroscopy. Med Eng Phys 30:693-699.
12. Ramos A, Bertemes-Filho P. (2011) Numerical sensitivity modeling for the detection of skin tumors by using tetrapolar probe. Electromagn Biol Med 30:235-245.
13. Brown BH, Tidy JA, Boston K, et al. (2000) Relation between tissue structure and imposed electrical current flow in cervical neoplasia. Lancet 355:892–895.
14. Brown BH, Milnes P, Abdul S, et al (2005) Detection of cervical intraepithelial neoplasia using impedance spectroscopy: a prospective study. Br J Obstet Gynaecol 112:802–806.
15. Gonzalez-Correa CA, Brown BH, Smallwood RH, et al (2003) Low frequency electrical bioimpedance for the detection of inflammation and dysplasia in Barrett's oesophagus Physiol Meas 24:291-296.
16. Gonzalez-Correa, CA. (2001) Endoscopic measurement of Electric Impedance Spectra and their Dependance on Tissue Properties in Barrett´s Oesophagus. Thesis. University of Sheffield. pp. 184, 82.
17. Atuma C, Strugala V, Allen A (2001) The adherent gastrointestinal mucus gel layer: thickness and physical state in vivo Am J Physiol Gastrointest Liver Physiol; 280: G922–G929.
18. Geselowitz D (1971). An application of electrocardiographic lead theory to impedance plethysmography. IEEE Trans Biomed Eng BME 18: 38-41
19. Grimmes S, Martinsen (2000) Bioimpedance & Bioelectricity Basics. Academic Press, NY. Chap. 5 Geometrical Analysis, 133-134.

Author: Edelberto Mulett-Vásquez
Institute: Universidad de Caldas
Street: Calle 25 # 48 – 57
City: Manizales
Country: Colombia
Email: emulettv@une.net.co

Correlation between Algometry and Electrical Bioimpedance in Subjects with and without Fibromyalgia

E. Colina-Gallo[1], C.A. González-Correa[2], and D.A. Miranda-Mercado[3]

[1] Universidad de Caldas, Department of Human Physical Activity, Manizales, Colombia
[2] Universidad de Caldas, Department of Basics Sciences, Manizales, Colombia
[3] Universidad Industrial de Santander, Department of Physics, CIMBIOS, Bucaramanga, Colombia

Abstract— **Fibromyalgia (FM) is a condition characterized by chronic widespread pain and generalized tenderness, accompanied by fatigue, sleep disturbances and cognitive difficulties. The diagnosis is merely clinical, as no confirmatory objective method is available so far. Several types of muscle abnormalities have been reported in FM, at tissue, cellular and subcellular level, that could eventually alter the resistivity of the muscle. In this paper, we used electrical impedance myography to detect changes in the passive electrical responses of muscle in a group of subjects with Fibromyalgia (*n*=24) and a control group without it (*n*=21). The overall apparent muscle resistivity obtained (2,08 Ω m) is well in the range reported in the literature for longitudinal measurements. We calculated and compared brachial apparent resistivity at the lowest frequency measured by the equipment (5 kHz) between both groups and we also compared resistivity against algometric results. No statistical significant differences were found between the two groups.**

Keywords— **Fibromyalgia, bio-electrical impedance, algometry, skeletal muscle resistivity.**

I. Introduction

FM is a condition characterized by chronic widespread pain, generalized tenderness, accompanied by fatigue, sleep disturbances and cognitive difficulties [1]. Muscle pain may depend on both increased peripheral and/or central nervous system response to painful or non-painful peripheral stimuli. In genetically predisposed individuals, this persistent input produces peripheral and central sensitization [2]. Several types of muscle abnormalities have been reported in FM [3], such as, low levels of phosphates [4], increased levels of lactate and pyruvate [5], reduced blood flow [6], increased cytokines [7], DNA fragmentation [8] and changes in the number and size of mitochondria [9]. The diagnosis of fibromyalgia is clinical and so far there are no objective confirmatory diagnostic methods [10].

Bioelectrical impedance analysis is a safe, non- invasive and objective method to assess architecture and function at tissue level [11]. Current applied to muscle travels either longitudinally or transversally, impedance being lower when it flows parallel to muscle fibres. Localized BIA it's a modification of the traditional BIA method and is used for obtaining information on the condition of specified muscle groups [12]. This approach allows determining the effective resistivity, i.e., resistance measured per length unit of the segment [12]. Rutkove et al [13] found that effective longitudinal resistivity was altered in patients with neuromuscular disease. In their work, this technique has been called electrical impedance myography (EIM), and it has proven useful for assessing changes in muscle composition and architecture in conditions such as atrophy, edema, reinnervation and fat deposition [14–17].

In this paper, we used a technique of linear multifrequency electrical impedance myography to detect changes in the passive electrical responses of muscle in a group of subjects with FM and a control group without it. We calculated and compared brachial apparent resistivity at the lowest frequency (5 kHz) measured by the equipment between both groups and we also compared resistivity against algometric results.

II. Materials and methods

A. Subjects

Twenty-four women (aged 31–63 y) with a clinical diagnosis of FM, and who had had symptoms for more than 1 year, were recruited from an educational program for patients with FM. As control group, we used data obtained from twenty-one healthy female controls enrolled from the university staff. Some background and anthropometric data of all subjects are given in Table 1.

Table 1 Physical characteristic of subject $\overline{X}$ (SD)

	Age (years)	Weight (kg)	Height (m)	BMI (kg/m^2)	% BF
FM (n=24)	48,5 (8,7)	67,7 (9,9)	1,56 (0,05)	27,8 (4,2)	34,8 (6,8)
Control (n=21)	52,4 (11,9))	67,3 (11,3)	1,55 (0,06)	28,1 (4,3)	35,3 (6,1)

F. Simini and P. Bertemes-Filho (eds.), *II Latin American Conference on Bioimpedance*,
IFMBE Proceedings 54,
DOI: 10.1007/978-981-287-928-8_19

B. *FM study variables*

FM diagnosis was confirmed or ruled out by applying the modification of the American College of Rheumatology (ACR) preliminary diagnostic criteria for fibromyalgia [10]. We also evaluated the number of tender points using a digital algometer WAGNER FDX (Greenwich USA) and considering the diagnosis of fibromyalgia when 11 or more tender points were present according to the 1990 ACR classification criteria [18]. The total tender point score (TTPS) was calculated as the sum of the pain threshold pressure at each point, divided by total tender points [19].

C. *Anthropometric measurements*

Weight and stature was measured with a digital measuring station SECA 284 and body fat percentage was measured with a TANITA analyzer (BC-418 model). In order to calculate the arm muscular area (AMA) [20] for the dominant side, the circumference of the midarm was measured with a cloth tape midway between the tip of the acromion and the olecranon process. Biceps and triceps skinfolds were measured using a SLIM GUIDE SKINFOLD CALIPER, taking a vertical fold midway between the tip of the acromion and the olecranon process. Skinfold readings were made in triplicate and the results were averaged.

D. *Bioimpedance measurements*

Arm segmental measurements of raw resistance (R) were performed at 50 frequencies in a log spectrum ranging from 5 kHz to 1 MHz, with a nominal current of 400μA RMS, and using a 4000B Bio-impedance spectrum analyzer system by XITRON technologies (San Diego, USA). Measurements were carried out by the same operator (ECG), made in triplicate and averaged. They were performed with the subject lying supine on a non-conducting surface and were completed within five minutes. Injecting electrodes were placed on the dorsal metacarpus and the dorsal metatarsus, at the dominant side of each individual. Voltage electrodes were placed on the ventral side of the dominant arm over the biceps brachii, 5 cm proximal and 5 cm distal to the midpoint between the tip of the acromion and the olecranon process. Adhesive electrodes (292-STE by IMPEDIMED) were used, and the sites where they were going to be placed were cleaned with alcohol swabs before their application.

E. *Estimation of apparent muscle resistivity*

Muscle resistivity (ρ_M) at 5 kHz was calculated using the formula applied to cylinders

$$\rho = R * A/L \, , \tag{1}$$

where R is the resistance offered by the muscle, L is the length of the cylinder and A is its cross-sectional area. Arm muscular area was calculated as described in [20], and the length corresponds to the distance between the voltage electrodes (0,10 m). In order to estimate the resistance offered by the muscle (R_M) at 5 kHz, we first considered the arm as four parallel resistances offered by dermis (R_D), fat (R_F), muscle (R_M) and bone (R_B). Total resistance (R_T) was taken as the value read by the equipment, and, for estimating the resistance of different compartments, we rearranged (1) to:

$$R = \rho * L/A \tag{2}.$$

We assumed resistivity values for dermis and fat as 4,5 and 30,0 Ω m, respectively, both in the ranges given in the literature [21,22], and length was 0,1m for both cases. In order to calculate the area corresponding to the dermis and the subcutaneous fat, we proceeded as follows: from the total area of the arm segment, we subtracted the area corresponding to a circle with radius equal to total radius minus 0,0015 m for the dermis. Fat area was calculated as total area of the arm minus the area estimated for the dermis minus the area corresponding to muscle and bone. The latter was obtained from a radius equal to total radius of the segment minus the average of biceps skinfold and triceps skinfold divided by 2.

F. *Data analysis*

A statistical analysis was performed using XLSTAT 2014. For assessing the statistical significance of the difference between mean resistances Student´s t-test were used.

III. Results

The application of the two diagnostic criteria allowed an adequate differentiation between the fibromyalgia and the control groups. As it can be seen in Table 2, the TTPS also separate them very well. The two groups were homogeneous in all physical characteristics including circumference of midarm, triceps and biceps skinfolds, and the areas estimated for the arm segment in consideration as well as in its compartments (see Table 3).

Table 2 Anthropometry and algometry data $\overline{X}$ (SD)

	Number of tender points	TTPS (kg/cm²)	Arm circumference (m)	Biceps skinfold (m)	Triceps skinfold (m)
FM (n=24)	16,0 (2,6)	2,40 (0,6)	0,30 (0,034)	0,012 (0,005)	0,026 (0,007)
Control (n=21)	5,4 (3,5)	4,06 (0,3)	0,30 (0,033)	0,012 (0,007)	0,025 (0,006)
P value	0,000	0,0	0,68	0,62	0,62

Table 3 Calculated areas (in m²) of the different arm compartments

	Arm	Dermis	Fat	Muscle + Bone	Muscle
FM (n=24)	0,0073 (0,0016)	0,0004 (0,0001)	0,0026 (0,0010)	0,0043 (0,0009)	0,0033 (0,0007)
Control (n=21)	0,0074 (0,0017)	0,0004 (0,0001)	0,0026 (0,0010)	0,0042 (0,0008)	0,0033 (0,0007)
P value	0,68	0,94	0,87	0,45	0,83

Data of arm segmental resistance at 5 kHz and 1 MHz, and specific muscular resistivity for biceps brachii at 5 kHz are given in Table 4 and Table 5. There is no statistically significant difference between mean values of both groups.

Table 4 Electrical resistance (in Ω) at 5 kHz of the different arm compartments

	Arm	Arm (1MHz)	Dermis	Fat	Muscle + bone
FM (n=24)	55,1 (9,3)	42,7 (7,2)	1032,9 (117,6)	1341,5 (550,1)	61,3 (10,8)
Control (n=21)	57,7 (8,8)	45,2 (6,7)	1018,7 (112,0)	1371,2 (562,8)	64,4 (10,1)
P value	0,34	0,24	0,68	0,86	0,31

Table 5 Calculated resistivity (ρ in Ω m) at 5 kHz of the different arm compartments

	Rho Arm	Rho Muscle + Bone	Rho Muscle
FM (n=24)	3,96 (0,59)	2,60 (0,33)	1,98 (0,31)
Control (n=21)	4,07 (0,72)	2,67 (0,45)	2,08 (0,42)
P value	0,6	0,62	0,32

In figures 1 and 2, it can be seen that algometry separates subjects with FM from those without it, while electrical bioimpedance does not.

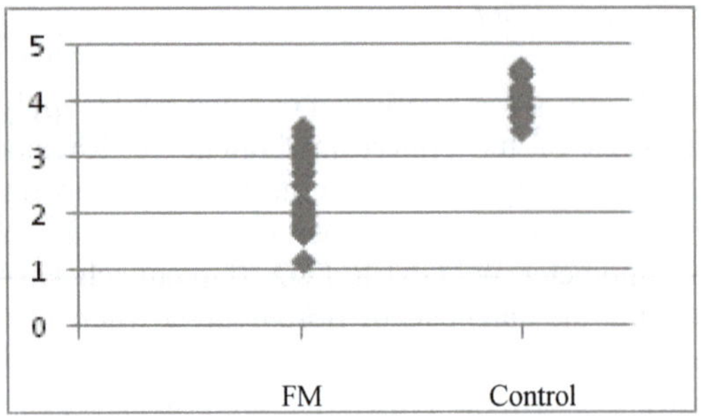

Fig 1 Total tender point score (TTPS)

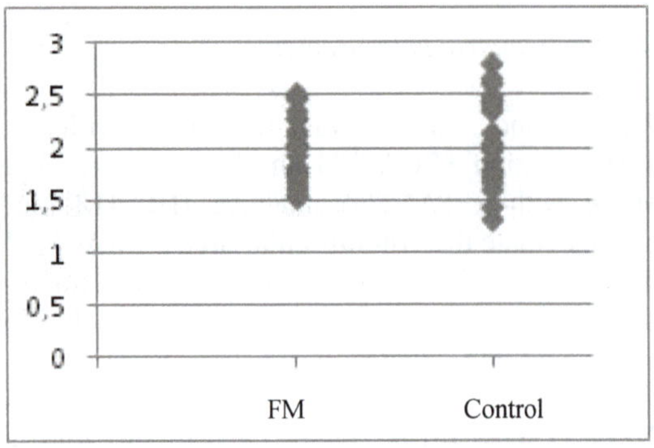

Fig 2 Calculate muscular resistivity at 5 kHz (Ω m

IV. Discussion

Our resistivity results are comparable with those reported in the literature [21,22]. In the meta-analysis by [22], for instance, the 95% confidence interval for skeletal muscle resistivity was 1,55-3,72 Ω m. [23] reported an average resistivity at zero frequency of five muscles for both sexes and all ages of 2 ± 0,19 ohm m.

Several mechanisms may explain changes in bioimpedance data in neuromuscular disease: resistance will increase with atrophy and loss of muscle fibers, fatty infiltration or connective tissue. Changes in intra and extra-cellular fluids or localized edema also impact the measured electrical impedance [14]. While some of these changes, for instance, muscle fiber damage and accumulation of pro-inflammatory substances, have been described in fibromyalgia, the data obtained in this study did not produce the expected differences in resistivity.

V. Conclusion

No statistically significant differences in resistivity of brachial muscle were found between subjects with fibromyalgia and those of the control group without FM.

However, it is necessary to consider various aspects: firstly, the small sample size, and, secondly, that, although muscular pain is generalized in FM, brachial muscle tenderness is less affected than periscapular or lumbar muscle tenderness. For this study, the choice of the brachial segment was due to technical accessibility and the availability of a formula to calculate muscle area.

Despite the fact of a negative result, we still consider convenient to carry out further studies involving a larger number of subjects and exploring other muscle groups like the trapezius.

Compliance with Ethical Requirements

All subjects gave their informed written consent in accordance with the guidelines of the Ethical Committee of the University of Caldas (Manizales, Colombia).

Conflict of Interest

The authors declare that they have no conflict of interest.

References

1. Buskila D (2009) Developments in the scientific and clinical understanding of fibromyalgia. Arthritis Res Ther 11:242 DOI 10.1186/ar2720
2. Bengtsson A (2002) The muscle in fibromyalgia. Rheumatology 41:721–4
3. Staud R (2011) Peripheral pain mechanisms in chronic widespread pain. Best Pr Res Clin Rheumatol 25(2):155–64 DOI 10.1016/j.berh.2010.01.010
4. Bengtsson A, Henriksson KG, Larsson J (1986) Reduce high-energy phosphate levels in the painful muscles of patients with primary fibromyalgia. Arthritis Rheum 29(7):817–21
5. Gerdle B, Söderberg K, Salvador Puigvert L et al. (2010) Increased interstitial concentrations of pyruvate and lactate in the trapezius muscle of patients with fibromyalgia: a microdialysis study. J Rehabil Med 42:679–87 DOI 10.2340/16501977-0581
6. McIver KL, Evans C, Kraus RM et al. (2006) NO-mediated alterations in skeletal muscle nutritive blood flow and lactate metabolism in fibromyalgia. Pain 120:161–9 DOI 10.1016/j.pain.2005.10.032
7. Wallace DJ, Linker-Israeli M, Hallegua D et al. (2001) Cytokines play an aetiopathogenetic role in fibromyalgia: a hypothesis and pilot study. Rheumatology 40:743–9 DOI 10.1093/rheumatology/40.7.743
8. Sprott H, Salemi S, Gay RE et al. (2004) Increased DNA fragmentation and ultrastructural changes in fibromyalgic muscle fibres. Ann Rheum Dis 63:245–51 DOI/10.1136/ard.2002.004762
9. Cordero MD, De Miguel M, Moreno Fernández AM et al. (2010) Mitochondrial dysfunction and mitophagy activation in blood mononuclear cells of fibromyalgia patients: implications in the pathogenesis of the disease. Arthritis Res Ther 12:R17 DOI 10.1186/ar2918
10. Wolfe F, Clauw DJ, Fitzcharles MA et al. (2011) Fibromyalgia criteria and severity scales for clinical and epidemiological studies: a modification of the ACR Preliminary Diagnostic Criteria for Fibromyalgia. J Rheumatol 38:1113–22 DOI 10.3899/jrheum.100594
11. Foster KR, Lukaski HC (1996) Whole-body impedance-what does it measure?. Am J Clin Nutr 64(Suppl):388S – 396S
12. Aaron R, Shiffman CA (2000) Using localized impedance measurements to study muscle changes in injury and disease. Ann N Y Acad Sci. 904:171–80
13. Rutkove SB, Aaron R, Shiffman CA (2002) Localized bioimpedance analysis in the evaluation of neuromuscular disease. Muscle Nerve 25:390–7 DOI 10.1002/mus.10048
14. Rutkove SB (2009) Electrical impedance myography: background, current state, and future directions. Muscle Nerve. 40(6):936–46 DOI 10.1002/mus.21362.Electrical
15. Rutkove SB, Esper GJ, Lee KS et al. (2005) Electrical impedance myography in the detection of radiculopathy. Muscle and Nerve 32:335–41 DOI 10.1002/mus.20377
16. Rutkove SB (2009) Electrical impedance myography as a biomarker for ALS. Lancet Neurol 8(3):226–7 DOI 10.1016/S1474-4422(09)70030-4
17. Rutkove SB, Darras B (2013) Electrical impedance myography for the assessment of children with muscular dystrophy: a preliminary study. J Phys Conf Ser 434(1) DOI 10.1088/1742-6596/434/1/012069
18. Wolfe F, Smythe HA, Yunus MB et al. (1990) The American College of Rheumatology 1990 criteria for the classification of fibromyalgia: report of the Multicenter Criteria Committee. Arthritis Rheum 33:160–72
19. Maquet D, Croisier JL, Demoulin C et al. (2004) Pressure pain thresholds of tender point sites in patients with fibromyalgia and in healthy controls. Eur J Pain 8:111–7 DOI 10.1016/S1090-3801(03)00082-X
20. Heymsfield SB, McManus C, Smith J et al. (1982) Anthropometric measurement of muscle mass: revised equations for calculating bone-free arm muscle area. Am J Clin Nutr 36:680–90
21. Miklavcic D, Pavselj N, Hart FX (2006) Electric properties of tissues. Wiley encyclopedia of biomedical engineering 1-12 John Wiley & Sons, Inc DOI 10.1.1.65.7364
22. Faes TJ, van der Meij HA, de Munck JC et al. (1999) The electric resistivity of human tissues (100 Hz-10 MHz): a meta-analysis of review studies. Physiol Meas 20:R1–10 DOI 10.1088/0967-3334/20/4/201
23. Shiffman CA, Rutkove SB (2013) Circuit modeling of the electrical impedance: part II normal subjects and system reproducibility. Physiol Meas 34(2):223–35 DOI 10.1088/0967/-3334/34/2/223

Author: Evelyn Colina Gallo
Institute: Universidad de Caldas
Street: Calle 25 #48-57
City: Manizales
Country: Colombia
Email: evelyn.colina@ucaldas.edu.co

Use of Bioimpedance Method to Quantify Changes in Left Ventricular Contractility in Experiments on Anesthetized Rats

V.V. Ermishkin[1,2], E.V. Lukoshkova[1], V.L. Lakomkin[1], A.A. Abramov[1], O.S. Tarasova[2,3], O.L. Vinogradova[2], and V.I. Kapelko[1]

[1] Russian Cardiology Scientific and Production Complex, Moscow, Russia
[2] Institute of Biomedical Problems of the Russian Academy of Sciences, Moscow, Russia
[3] M.V. Lomonosov Moscow State University, Moscow, Russia

Abstract — **The impedance method of pre-ejection period (PEP) evaluation had been adapted for a small animal study and validated with a Millar-type pressure sensor inserted into aorta near the aortic valve or in the left ventricle. A strong correlation (*r*=0.95) was found between the PEP values determined by the impedance and micromanometer methods. The difference between the PEP estimates was about 5 ms. A significant correlation (*r*=0.75-0.78) was also found between the PEP and invasive indices of left ventricular contractility, $dP/dt|_{max}$ and $dP/dt|_{max}/P$. It was observed for a variety of the baseline states added with those during dopamine or ketamine injections. The contractility changes induced by a number of substances were evaluated by the invasive and impedance methods, which showed very similar time course of the responses. Meanwhile, the diastolic filling of the heart or changes in the arterial pressure during the tilt test or caused by a hypotensive agent may also affect PEP. Our findings suggest that impedance method can provide an accurate estimation of PEP changes in animal experiments. So, it can be a useful tool for quantifying potential inotropic effects of pharmacological agents. However, it should be also taken into consideration the dependency of PEP on the preload and afterload conditions, especially, when vasodilation effects persist.**

Keywords — **cardiac contractility, impedance cardiography, pre-ejection period, dP/dt, anesthetized rats.**

I. Introduction

Contractility is a fundamental property of cardiac ventricular myocytes that eventually determine the maximum power of the pumping heart. Its assessment is of great clinical importance. The maximum rate of ventricular pressure rise ($dP/dt|_{max}$) remains the most widely used index of cardiac contractility in animal studies [1]. However, the use of invasive methods has many restrictions. So, in human studies, the noninvasive methods of cardiac contractility measurements became most popular.

One approach is based on evaluation of the pre-ejection period (PEP), which can be obtained by simultaneous recording of the thoracic impedance cardiogram (ICG) and electrocardiogram (ECG). Traditionally, PEP is defined as the interval from the onset of left ventricular depolarization (Q-wave onset in the ECG) to the opening of the aortic valve [2], reflected by the B-point in the ICG signal [3, 4].

This short systolic time interval is highly sensitive to sympathetic activation and other inotropic influences [2-4], and may be considered as a method of choice to monitor changes in cardiac sympathetic activity non-invasively [4]. However, it should be taken into account that PEP duration also depends on the left ventricular (LV) end-diastolic volume and diastolic pressure in the ascending aorta [3-5].

Now it is generally accepted that the B-point in ICG dZ/dt is the best approximation of PEP end (or the onset of ejection) [4]. However, in practice, an accurate and unequivocal determination of PEP can be often problematic due to complexity and variability of the dZ/dt waveform [3, 5-8]. The problem of accurate determination of the PEP duration became even more pronounced when applied to small laboratory animals having higher heart rates, rather noisy ICG signals, and enhanced respiratory impedance component. In this study we used a more robust surrogate method of PEP evaluation, taking R peak in ECG as the start point and the moment when the second derivative of the ΔZ signal attains the maximum ($d^2Z/dt^2|_{max}$) for the end of PEP [5]. We tested this method for quantification of changes in LV contractility in animal experiments during tilt test and several pharmacological interventions and compared the results with the synchronously obtained direct catheter measurements.

II. Methods

Experiments were carried out on male Wistar rats (400-450 g) anesthetized with Ketamine (100 mg/kg). ICG signals from the thoracic region (ΔZ) and ECG in lead II were recorded simultaneously with an impedance cardiograph RPKA-2-01 (*MEDACC*, Moscow) having a modified filter bandwidth of 0.1 to 150 Hz. The current electrodes (steel needles) were fixed at the head and a hind limb and two voltage electrodes were inserted at the top and the lower edges of the sternum. A catheter with a 1.6 F Scisense micromanometer pressure sensor was introduced via the carotid artery and connected to a Hugo Sachs Elektronik strain-gauge amplifier. It was used

F. Simini and P. Bertemes-Filho (eds.), *II Latin American Conference on Bioimpedance*,
IFMBE Proceedings 54,
DOI: 10.1007/978-981-287-928-8_20

for high fidelity recordings of the LV pressure (P_{LV}) or the aortic pressure (P_{Ao}). Systemic blood pressure was controlled using a solution-filled catheter positioned in the abdominal aorta.

The signals were continuously sampled at the rate of 1 kHz with a 16-bit analogue-to-digital converter (National Instruments, Texas) and stored on a hard disk. Every 5 s the waveforms of the ICG and pressure signals recorded for 20-30 successive cardiac cycles were synchronized by ECG R peak and averaged. Then the first and the second derivatives of these ensemble-averaged signals were calculated point-by-point using a 3-rd order polynomial fitting over 5-7 points by Savitzky-Golay method [8]. All analyzed signals are shown in Fig. 1.

The value of PEP_Z for each averaged ΔZ wave was calculated as the interval between the R and $d^2Z/dt^2|_{max}$ points. In a similar way, the values of PEP_{Ao}, corresponding to the start of systolic pressure waves registered by the Scisense sensor, which was arranged in the aorta near the aortic valve, were calculated. Afterwards, the pressure sensor was moved into the LV, and the invasive contractility indices, such as $dP/dt|_{max}$ or $dP/dt|_{max}/P$ (with P taken at the point where dP/dt reaches the maximum), were also calculated. The latter index was proposed to minimize the load dependency of dP/dt [8]. The amplitudes of $dZ/dt|_{max}$ and $d^2Z/dt^2|_{max}$ and the systolic interval RZ, from the R to the $dZ/dt|_{max}$ points (see [7]), were also determined.

The invasive and non-invasive estimates of LV contractility were compared under various experimental conditions including the baseline states, tilt test (60°, 3 min), infusion of

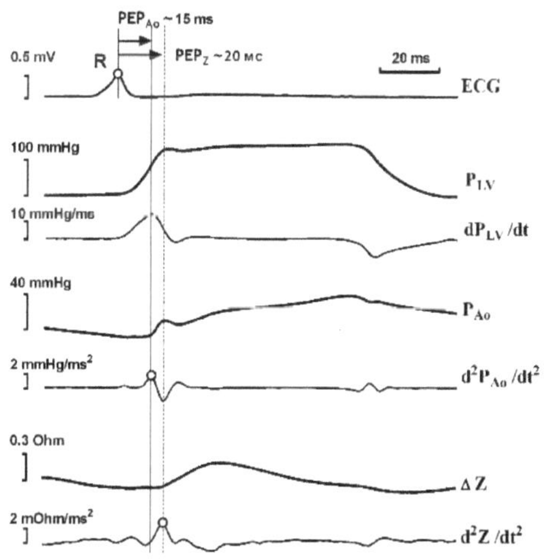

Fig. 1 Ensemble-averaged original signals (ECG, P_{LV}, P_{Ao}, ΔZ) and the calculated (dP_{LV}/dt, d^2P_{Ao}/dt^2, d^2Z/dt^2) signals. The pressure in the LV and aorta was recorded with the same catheter-tip manometer during different cardiac cycles with similar durations. For convenient presentation of the related blood volume changes, ΔZ curves and their derivatives, are inverted, i.e., the positive deflections of the curves correspond to the decrease in impedance.

Dopamine (10-20 μg/kg/min) or Dobutamine (1-5 μg/kg/min), additional injection of Ketamine (10-20 mg/kg). In a few rats, effects of Sodium Nitroprusside (SNP, 1-10 μg/kg/min), or autonomic blockade with Methylatropine (1 mg/kg) and Atenolol (2 mg/kg) were also tested. All substances were introduced via the jugular vein catheter.

Data acquisition and processing was performed using the custom-made original software designed in LabVIEW (National Instruments, Texas).

III. Results

In a series of experiments with the pressure sensor set in the aorta (13 rats), the values of PEP_Z and PEP_{Ao} estimates were compared. The results of a typical experiment shown in Fig. 2 demonstrate similar responses of PEP_Z and PEP_{Ao} to increasing doses of Dopamine which mimicked the effects of sympathetic activation on the heart. Parallelism between the PEP_Z and PEP_{Ao} changes was evident not only in the large-scale fragments (plot ***a***) but also at the smaller time intervals (insertion underneath). The regression plot (***b***) exhibited a strong correlation between these two estimates of PEP. Bland-Altman analysis (***c***) also showed a good consistency of PEP_Z and PEP_{Ao} with a rather stable shift (4-5 ms) between them.

The minimal correlation coefficients were observed during the tilt test (r=0.95±0.05, n=10) while the highest – for Dobutamine infusions (r=0.99±0.01, n=12).

In other series of experiments, with the micromanometer placed inside the left ventricle, the pressure-derived indices of contractility were compared with PEP_Z and other relevant

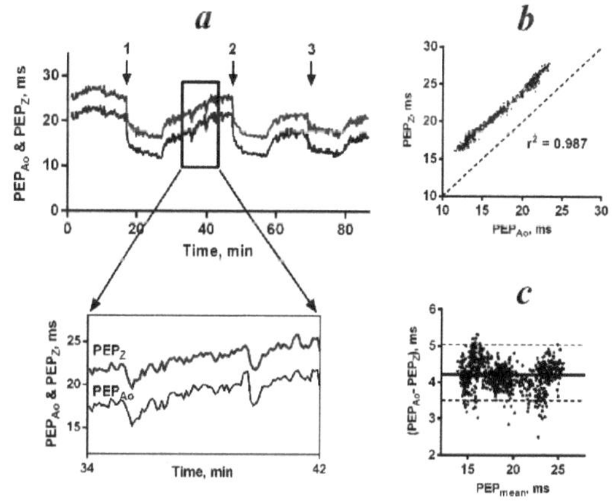

Fig. 2 Comparison of PEP values obtained with the impedance (PEP_Z) and catheter (PEP_{Ao}) methods. ***a*** – trends of calculated PEP_Z and PEP_{Ao} values; the arrow heads 1, 2, and 3 indicate the start of Dobutamine infusion with the rate of 20, 15 and 10 μg/kg/min; ***b*** – regression plot and ***c*** – Bland-Altman plot for these values.

impedance parameters (Table 1). The baseline measurements carried out in 60 rats before any tests demonstrated that dP/dt indices correlated well with the initial systolic time intervals PEP_Z and RZ but not with the amplitudes of $dZ/dt|_{max}$ and $d^2Z/dt^2|_{max}$.

Table 1 Correlation (Spearman) between impedance and pressure catheter parameters in rats anesthetized with Ketamine, baseline records (n=60)

Parameters	$dP_{LV}/dt|_{max}$	$dP_{LV}/dt|_{max}/dP_{LV}$	HR
PEP_Z	-0.75	-0.78	-0.57
RZ	-0.76	-0.77	-0.60
$1/PEP_Z$	0.71	0.76	0.53
$dZ/dt|_{max}$	0.18	0.14	0.22
$d^2Z/dt^2|_{max}$	0.16	0.16	0.17

Taking into account the delay of PEP_Z against PEP_{Ao} and the opposite direction of changes in LV contractility and the duration of PEP (the higher contractility, the shorter is PEP), we proposed a new index $1/(PEP_Z\text{-}5)$ which lacks the above mentioned shortcomings. Fig. 3 shows a correlation of $1/(PEP_Z\text{-}5)$ with an invasive LV pressure index $dP/dt|_{max}/P$ found in a big pool of the baseline states (60 rats) merged with the states corresponding to peak inotropic effects of Dopamine or additional Ketamine (n=40 points).

Fig. 4 displays examples of LV contractility responses during four different tests: β1-agonist Dobutamine infusion, autonomic blockade, tilt test and vasodilator SNP infusion. In the first three tests the kinetics of changes in the non-invasive index $1/(PEP_Z\text{-}5)$ was quite similar to the changes in $dP/dt|_{max}$, but the vasodilator SNP sometimes induced the inverse responses of the PEP-based index and $dP/dt|_{max}$. It is likely that a decrease in contractility reflected by the $dP/dt|_{max}$ resulted from the lower diastolic filling of the ventricle due to blood deposition in the dilated vessels. At the same time, a drop in the aortic diastolic pressure produced by the vasodilator could lead to an earlier opening of the aortic valve, i.e. to a shorter PEP.

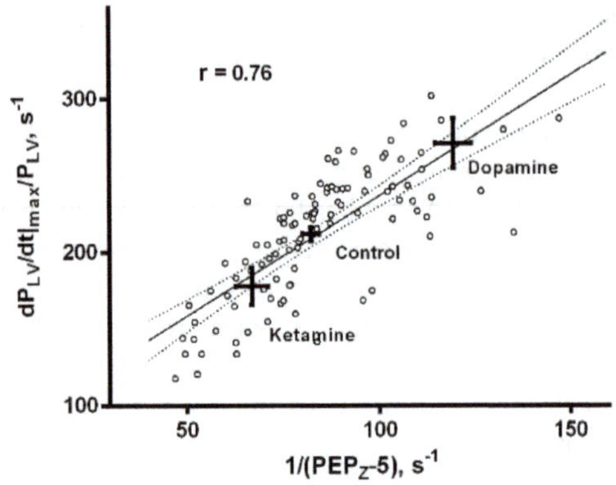

Fig. 3 Correlation (by Pearson) between the bioimpedance and invasive LV pressure contractility indices. 100 pairs correspond to the baseline measurements in 60 rats and to peak responses to Dopamine or additional injection of Ketamine. The crosses represent the average values (M±SE) of the indices at these three states.

IV. Discussion

ICG had been elaborated as a method for application in humans. So, the major fiducial points for timing the onset and end of ejection in ICG were found by matching it with other non-invasively recorded signals, predominantly under the rest conditions. Analysis of the synchronously recorded thoracic impedance and LV or aortic pressure signals in rats under control of hemodynamic parameters provides an

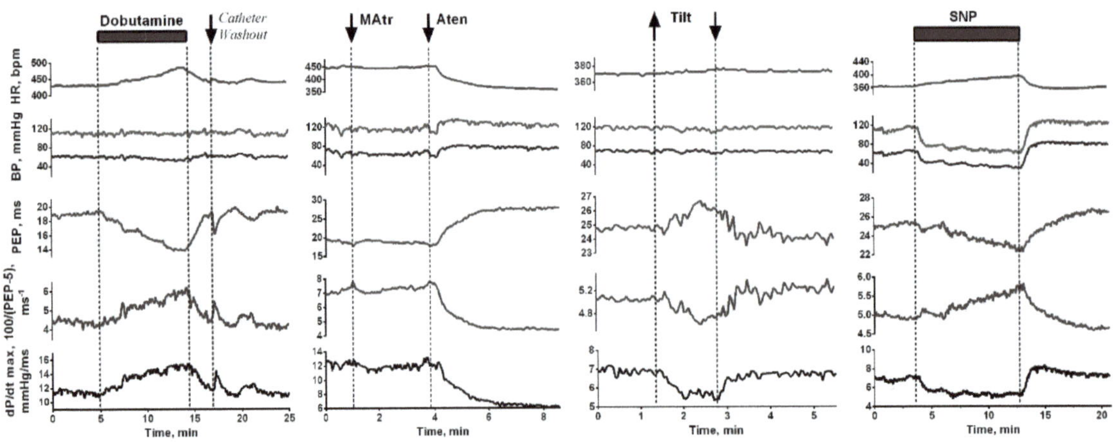

Fig. 4 LV contractility responses in anesthetized rats: to β1-agonist, autonomic blockade, tilt test, and vasodilator agent (given from the left to the right). For details see the text.

alternative view on the phenomena associated with the particular events in the ICG signal. For example, a small pre-ejection peak can be seen in the traces of d^2Z/dt^2 and d^2P_{Ao}/dt^2 which appeared near the maximum of R peak in ECG (Fig 1). The nature of such pre-ejection waves is not clear [8], however, the similarity of these waves in both the traces probably points to their common origin.

The current work has focused at elucidating two aspects of ICG method employment for LV contractility assessment: 1) the accuracy of PEP determination by the impedance method and 2) a relation between PEP and invasive indices of contractility.

An excellent agreement of PEP_Z and PEP_{Ao} trends observed in our experiments confirmed the earlier findings of Siegel et al [10], who also showed a strong correlation between the intervals from R peak in the ECG to the inflection point in dZ/dt (i.e. PEP_Z) and to the LV $dP/dt|_{max}$ point (analog of PEP_{Ao}). It is likely that a stronger correlation observed in our study resulted from the use of similar algorithm of identification of the onset of systolic wave for both the pressure and impedance signals. The idea of the proposed d^2Z/dt^2 method was to avoid problems associated with searching the B-point and to find another event in ICG curve which also relates to the LV ejection onset but can be identified with a computer algorithm more easily and reliably. This approach is an immediate relative of ISTI-method used by Meijer et al [7]. Previously, similar algorithms were successfully employed for studying ISTI and PEP responses in humans [5, 7]. These approaches, however, are sometimes put to criticism [4]. Nonetheless, our experiments demonstrated an excellent consistency between PEP_Z and PEP_{Ao} estimates during most of pharmacological interventions in anesthetized rats.

A new index $1/(PEP_Z\text{-}5)$ adequately responded to experimentally induced changes in the inotropic state and correlated with changes in $dP/dt|_{max}$. This result agrees with the data obtained in the conscious telemeterised rats [1] that the interval from the Q wave in the ECG to the start of the blood pressure wave in the abdominal aorta was inversely related to the LV $dP/dt|_{max}$.

V. Conclusions

Simultaneous recording of the thoracic impedance and LV pressure may help to elucidate the phenomena associated with the ICG signal and also for testing the capability of the ICG method to assess cardiac contractility.

Our results suggest that bioimpedance method can be useful in testing or controlling the inotropic effects of the injected substances if the more trustful but expensive and sophisticated techniques cannot be employed. The bioimpedance method can be successfully applied for toxicological screening or to control the animal state under anesthesia.

Acknowledgment

The study was supported by the Russian Academy of Sciences, the Program "Fundamental Sciences for Medicine - 2014", and by the Russian Foundation for Basic Research (grants 12-04-01104-a and 15-04-06571-a).

Conflict of Interest

The authors declare that they have no conflict of interest.

References

1. Adeyemi O, Roberts S, Harris J et al. (2009) QA interval as an indirect measure of cardiac contractility in the conscious telemeterised rat: model optimisation and evaluation. J Pharmacol Toxicol Methods 60 (2):159–166
2. Lewis RP, Leighton RF, Forester WF, Weissler AM (1974) Systolic time intervals. In: Non-invasive cardiology (AM Weissler, ed). Grune & Stratton, NY, 1974, pp. 301–368
3. Sherwood A, Allen MT, Fahrenberg J et al. (1990) Methodological guidelines for impedance cardiography. Psychophysiology 27:1–23
4. van Lien R, Schutte NM, Meijer JH, de Geus EJC (2013) Estimated preejection period (PEP) based on the detection of the R-wave and dZ/dt-min peaks does not adequately reflect the actual PEP across a wide range of laboratory and ambulatory conditions. Int J Psychophysiol 87:60–69
5. Ermishkin VV, Lukoshkova EV, Bersenev EY et al. (2007) Beat-by-beat changes in pre-ejection period during functional tests evaluated by impedance aortography: a step to a left ventricular contractility monitoring. IFMBE Proc 17:655–658
6. Lozano DL, Norman G, Knox D et al (2007) Where to B in dZ/dt. Psychophysiology 44:113–119
7. Meijer JH, Boesveldt S, Elbertse E et al. (2008) Method to measure autonomic control of cardiac function using time interval parameters from impedance cardiography. Physiol Meas 29:383–391
8. Ermishkin VA, Kolesnikov VA, Lukoshkova EV (2014) Age-dependent and 'pathologic' changes in ICG waveforms resulting from superposition of pre-ejection and ejection waves Physiol Meas 35:943–963
9. Veragut UP, Krayenbuhl HP (1965) Estimation and quantification of myocardial contractility in the closed chest dog. Cardiologia 479 (2):96–112
10. Siegel JH, Fabian M, Lankau C et al. (1970) Clinical and experimental use of thoracic impedance plethysmography in quantifying myocardial contractility. Surgery 67:907–917

Address of the corresponding author:

Author: Ermishkin Vladimir V.
Institute: Russian Cardiology Scientific & Production Complex
Street: 3rd Cherepkovskaya street, 15a
City: Moscow
Country: Russia
Email: v.v.erm@mail.ru

Evaluation of the Heath-Carter Somatotype Revisited: New Bioimpedance Equations for Children and Adolescents

A.V. Anisimova[1], E.Z. Godina[1], D.V. Nikolaev[2,4], and S.G. Rudnev[3,4]

[1] Institute and Museum of Anthropology, Moscow State University, Moscow, Russia
[2] Scientific Research Centre 'Medas', Moscow, Russia
[3] Institute of Numerical Mathematics, Russian Academy of Sciences, Moscow, Russia
[4] Central Research Institute for Health Organization and Informatics, Moscow, Russia

Abstract— **Based on the results of the cross-sectional anthropological study of 2364 Russian children and adolescents aged 7-17 years, we suggest simple prediction formulae for automated bioimpedance-based evaluation of endomorphy and mesomorphy components of the Heath-Carter somatotype: $ENDO_{BIA} = 0.5282\times FMi + 0.2580\times BMI - 0.04822\times BM - 1.881$ (r^2=0.81, SEE=0.65); $MESO_{BIA} = 0.3651\times FFMi + 0.42765\times BMI - 0.09323\times BM - 4.803$ (r^2=0.81, SEE=0.54), where BMI, FMi and FFMi are, respectively, the body mass, fat mass and fat-free mass indices (kg/m^2), and BM is the body mass (kg). In addition, in order to avoid using indirect bioimpedance body composition estimates, alternative formulae are constructed based only on directly measured rather than estimated bioimpedance data: $ENDO_{BIA} = -3224.7/R + 0.63867\times BMI - 0.04162\times BM - 2.195$ (r^2=0.81, SEE=0.65); $MESO_{BIA} = 2195.4/R + 0.52966\times BMI - 0.09740\times BM - 4.5522$ (r^2=0.81, SEE=0.54), where R is the whole-body electrical resistance (Ohm) at a frequency of 50 kHz. These formulae can be used for the specified age range regardless of sex and, due to relatively high proportion of the explained variance, are suitable for individual typology.**

Keywords— **Somatotype, Heath-Carter typology, bioelectrical impedance analysis, the whole-body electrical resistance, fat mass index, fat-free mass index, prediction formulae.**

I. Introduction

The terms somatotyping and constitution study are generally used for the designation of one of the methods for the analysis and classification of body physique [1-5]. The Heath-Carter anthropometric somatotype [6] that was suggested as the development of the classical Sheldon's photoscopic scheme of the assessment of body physique [1], is one of the commonly used methods and still of important significance for anthropology and sports science [7-9].

The Heath-Carter somatotype represent an ordered set of three numbers: endomorphy (which is regarded as a relative body fatness), mesomorphy (a measure of musculoskeletal development), and ectomorphy (relative linearity of physique). Software for the Heath-Carter anthropometric somatotype calculation and management is available [6,10,11]. With this, the assessment of the Heath-Carter somatotype is not always possible because a significant number of anthropometric measurements is needed which require considerable expertise.

Classical studies revealed significant relationships of the Heath-Carter endomorphy component with percent body fat both in adults and children [12,13], and of the mesomorphy component with lean body mass in adults [12], whereas in children the mesomorphy showed little association with lean body mass alone or in combination with height and weight [13]. In their study of 260 adolescent boys aged 16 to 18 years, T. Nawarycz and L. Ostrowska-Nawarycz suggested an approach for the computerized analysis of the first and the second components of the Heath-Carter somatotype using bioimpedance analysis [14], now the most promising simple and easy to use method of body composition assessment [15]. Their regression equation for the endomorphy component was based on the bioimpedance percentage body fat (%BF), whereas the mesomorphy component was determined using body height, widths of humerus and femur epiphyses, circumferences of the upper arm and the calf, and the BIA %BF instead of skinfold data [14]. So, the formula for the second component of the somatotype included a number of parameters not routinely measured within the standard procedure of bioimpedance measurements.

Our aim was to re-analyse the relationships between the Heath-Carter somatotype and body composition and to develop prediction formulae for automated bioimpedance-based evaluation of the somatotype in children and adolescents suitable for use in a wide range of age in both sexes.

II. Subjects and Methods

Anthropometry was performed in 2364 apparently healthy children and adolescents of the Russian ethnicity (1450 boys and 914 girls) aged 7-17 years using standard measurement protocol adopted at the Institute and Museum of Anthropology of the Lomonosov Moscow State University as described in [16]. The data were collected cross-sectionally in 2005-2013 at schools of Moscow (n=1456), Arkhangelsk (n=357), and Arkhangelsk region (n=551).

F. Simini and P. Bertemes-Filho (eds.), *II Latin American Conference on Bioimpedance*,
IFMBE Proceedings 54,
DOI: 10.1007/978-981-287-928-8_21

Standing height (Ht) was accessed by the GPM (Martin type) anthropometer. Body mass (BM) was measured on a digital scale to the nearest 100 g. Body mass index (BMI) was calculated as the BM relative to Ht squared (kg/m^2). Calf, triceps, subscapular and supraspinale skinfold thicknesses were measured on the right side of the body using the GPM (Harpenden type) skinfold caliper to the nearest 0.1 mm. Femur and humerus biepicondylar breadths, as well as arm and calf girths were measured using appropriate instrumentation to the nearest 0.5 mm. Endomorphy (Endo), mesomorphy (Meso) and ectomorphy (Ecto) components of the Heath-Carter anthropometric somatotype were determined based on the above mentioned quantities using conventional formulae as described in [6].

The whole-body impedance was measured in a supine position on the right side of the body according to a conventional tetrapolar measurement scheme by the bioimpedance analyzer ABC-01 'Medas' (SRC Medas, Moscow, Russia) at a frequency of 50 kHz using disposable Ag/AgCl Schiller bioadhesive electrodes.

Fat-free mass (FFM) was accessed using Houtkooper equation [17]: FFM = $0.61 \times (Ht^2/R) + 0.25 \times BM + 1.31$, where Ht is the standing height (cm), R is the whole-body electrical resistance (Ohm), and BM is the body mass (kg). Fat mass (FM) was obtained as the difference between BM and FFM. Similarly to BMI, fat-free mass index (FFMi) and fat mass index (FMi) were calculated as the ratio of FFM (kg) and FM (kg), respectively, to height squared (m^2).

All statistical analyses were performed using Minitab 17 and MS Excel 2007 software packages.

III. Results

Basic anthropometric characteristics of the study group are shown in Table 1, with (*) showing a statistically significant differences ($p<0.05$) between boys and girls for a given age.

Table 1 Height, weight and BMI of the study group according to age and sex, mean (SD)

Age	Body height, cm		Body mass, kg		BMI, kg/m^2	
	Boys	Girls	Boys	Girls	Boys	Girls
7	124.3 (6.7)	124.9 (6.8)	25.8 (4.9)	25.4 (5.2)	16.6 (2.0)	16.2 (2.3)
8	129.0 (6.3)	127.9 (5.8)	28.2 (5.3)	27.1 (5.0)	16.8 (2.0)	16.5 (2.2)
9	134.9 (6.0)	133.9 (5.8)	31.9 (6.3)	30.9 (6.1)	17.4 (2.7)	17.1 (2.6)
10	139.6 (5.6)	138.4 (7.1)	34.9 (6.4)	33.0 (7.1)	17.8 (2.6)	17.0 (2.3)
11	145.5 (8.0)	146.2 (7.9)	40.3 (10.0)	39.7 (10.1)	18.8 (3.4)	18.3 (3.2)
12	151.6 (7.1)	153.4 (8.2)	44.5 (8.8)	44.2 (11.1)	19.2 (3.0)	18.6 (3.4)
13	158.3 (8.8)	157.4 (7.6)	50.4 (11.4)	49.5 (11.5)	19.9 (3.1)	19.8 (3.6)
14	165.2 (9.5)*	161.6 (6.8)	56.4 (11.3)*	53.2 (10.3)	20.6 (3.0)	20.3 (3.2)
15	171.0 (8.4)*	162.3 (6.2)	61.8 (13.3)*	54.9 (8.5)	21.0 (3.6)	20.8 (2.9)
16	173.7 (7.2)*	164.6 (6.1)	65.3 (12.6)*	56.2 (7.5)	21.6 (3.3)	20.8 (2.7)
17	175.2 (6.5)*	162.4 (6.8)	66.4 (9.8)*	55.7 (8.0)	21.6 (2.7)	21.1 (2.5)

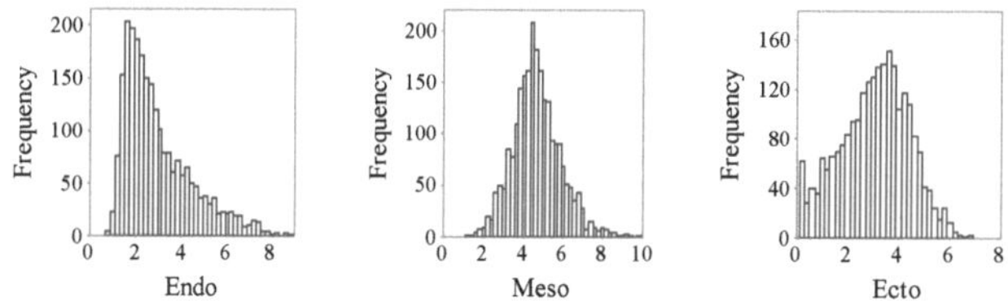

Fig. 1 The histograms of endo-, meso-, and ectomorphy components distributions of the Heath-Carter somatotype in the study group (n=2364)

Our data showed unimodal distributions of the Endo and Meso components of the somatotype in the study group having a pronounced positive skewness and kurtosis, respectively (see Fig. 1). The distribution of the Ecto component was also, largely, unimodal with a small additional peak at the value of 0.1 reflecting cumulative number of children on the left tail of the distribution, i.e. with zero or negative calculated values of the ectomorphy. The median somatotype of our study group was 2.5-4.5-3.2 that can be described as ectomorphic mesomorph according to Carter and Heath typology [6].

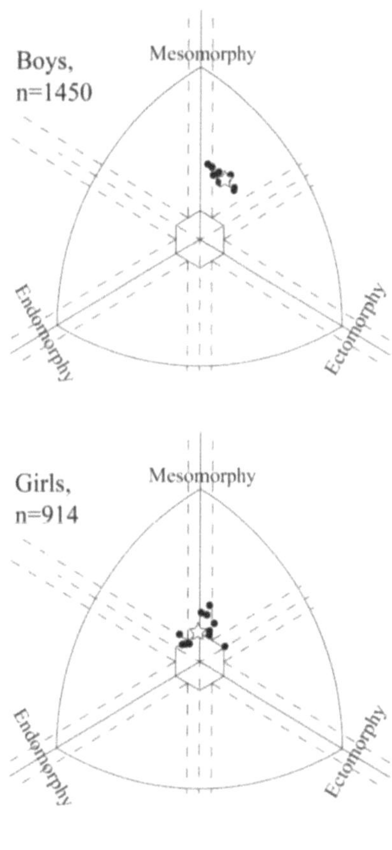

Fig. 2 The Heath-Carter somatocharts of the study group according to age and sex. Black circles show the median somatotypes for certain age (years); white star indicates the overall median somatotype

Table 2 The Heath-Carter somatotype of the study group according to age and sex

Age, years	Boys				Girls			
	n	Endo	Meso	Ecto	n	Endo	Meso	Ecto
7	50	2.2	5.0	2.5	47	2.3	4.5	2.7
8	86	2.2	5.0	2.7	94	2.6	4.4	2.9
9	79	2.2	4.9	2.9	82	2.9	4.6	2.9
10	90	2.3	4.7	3.1	43	2.8	4.4	3.4
11	103	2.3	5.1	3.1	66	2.9	4.2	3.3
12	118	2.5	5.0	3.0	97	2.7	3.8	3.7
13	152	2.4	4.9	3.2	100	2.9	4.0	3.3
14	191	2.1	4.8	3.3	98	3.4	3.9	3.0
15	221	2.1	4.5	3.5	110	3.6	3.9	3.0
16	217	2.0	4.9	3.3	103	3.6	3.9	2.9
17	143	2.0	4.5	3.4	74	3.6	4.1	2.7

In boys, our cross-sectional data showed the age trend from balanced mesomorph to ectomorphic mesomorph category (see Fig. 2 and Table 2), with the overall median ectomorphic mesomorph somatotype 2.2-4.8-3.2. The studied group of girls showed a more complex pattern of change, from balanced mesomorph to ectomorphic mesomorph and, then, through central phenotype, to endomorphic mesomorph category at the age of 17 thus reflecting adiposity traits in the somatic growth and sexual maturation. The overall somatotype of our girls was 3.1-4.2-3.1, or balanced mesomorph.

Table 3 Pearson's correlations between the Heath-Carter somatotype components and the bioimpedance body composition parameters in boys and girls (upper right and lower left parts of the table, respectively)

	Endo	Meso	Ecto	BM	BMI	FM	FMi	%FM	FFM	FFMi
Endo	•	0.69	-0.78	0.34	0.70	0.74	0.87	0.80	0.14	0.35
Meso	0.65	•	-0.89	0.31	0.72	0.53	0.64	0.49	0.19	0.57
Ecto	-0.80	-0.87	•	-0.24	-0.70	-0.57	-0.73	-0.64	-0.09	-0.47
BM	0.57	0.24	-0.41	•	0.85	0.77	0.53	0.30	0.96	0.86
BMI	0.80	0.63	-0.78	0.88	•	0.88	0.81	0.58	0.72	0.86
FM	0.73	0.38	-0.55	0.92	0.91	•	0.93	0.81	0.56	0.56
FMi	0.85	0.56	-0.72	0.79	0.92	0.95	•	0.94	0.28	0.39
%FM	0.79	0.40	-0.61	0.66	0.77	0.88	0.94	•	0.04	0.09
FFM	0.42	0.12	-0.26	0.97	0.78	0.79	0.62	0.48	•	0.87
FFMi	0.62	0.60	-0.69	0.81	0.90	0.70	0.67	0.43	0.81	•

The correlations of the Heath-Carter somatotype components Endo, Meso and Ecto in boys and girls with the indices of fat- and fat-free mass (FMi, FFMi) were higher as compared to absolute FM and FFM values or the %FM (Table 3). In this regard, we proposed the following simple prediction formulae for the bioimpedance evaluation of the Endo and Meso components of the somatotype:

$$ENDO_{BIA} = 0.5282\times FMi + 0.2580\times BMI - 0.04822\times BM - 1.881 \quad (r^2=0.81, SEE=0.65) \qquad (1)$$

$$MESO_{BIA} = 0.3651\times FFMi + 0.42765\times BMI - 0.09323\times BM - 4.803 \quad (r^2=0.81, SEE=0.54) \qquad (2)$$

All the components of the regression formulae (1) and (2) were essential (see Tables 4 and 5) with the regression lines for the residuals not significantly different from zero (Fig. 3).

Table 4 Contribution and order of entry of predictor variables to the regression model (1) for the endomorphy component of the Heath-Carter somatotype

Predictor variables	r^2	SEE	p
FMi	0.76	0.75	<0.001
BM	0.78	0.72	<0.001
BMI	0.81	0.65	<0.001

r^2 is the proportion of explained variance; SEE is the standard error of the model; p is the significance of contribution of the respective parameter to the stepwise multiple regression model

Table 5 Contribution and order of entry of predictor variables to the regression model (2) for the mesomorphy component of the Heath-Carter somatotype

Predictor variables	r^2	SEE	p
BMI	0.47	0.89	<0.001
BM	0.71	0.66	<0.001
FFMi	0.81	0.54	<0.001

r^2 is the proportion of explained variance; SEE is the standard error of the model; p is the significance of contribution of the respective parameter to the stepwise multiple regression model

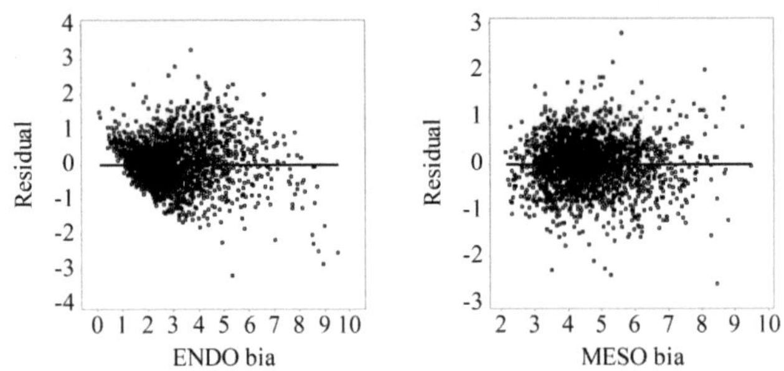

Fig. 3 The residuals and the respective regression lines for endomorphy and mesomorphy estimates of the Heath-Carter somatotype

One can note, due to mutual dependence of the FFM on the impedance index Ht^2/R, that the FFMi, as a ratio of FFM to Ht^2, should strongly correlate with the inverse value of the electrical resistance R. With this idea, in order to avoid using population-specific body composition equations, we constructed the alternative formulae for the evaluation of the Heath-Carter somatotype relying solely on measurements of height, weight, and the electric resistance:

$$ENDO_{BIA} = -3224.7/R + 0.63867\times BMI - 0.04162\times BM - 2.195 \quad (r^2=0.81, SEE=0.65) \qquad (3)$$

$$MESO_{BIA} = 2195.4/R + 0.52966\times BMI - 0.09740\times BM - 4.5522 \quad (r^2=0.81, SEE=0.54) \qquad (4)$$

These formulae are similar to Eqs. (1)-(2) in structure, have the same accuracy of the response variables approximation, and take an advantage of using only directly measured rather than estimated bioimpedance data. The relatively high values of the proportion of explained variance r^2 enable the use of these formulae for individual typology. Given that the ectomorphy, i.e., the third component of the somatotype, is calculated directly on patient's height and weight [6], we, thus, obtain an opportunity for automated bioimpedance-based evaluation of the overall Heath-Carter somatotype in children and adolescents. The respective algorithm was embedded in the current version of the ABC-01 'Medas' bioimpedance meter software.

IV. Conclusions

The assessment of body composition and somatotyping represent two different, but correlated, ways of describing human physique and structure. The Heath-Carter anthropometric somatotype [6] is one of commonly used methods of somatotyping and still of important significance for anthropology and sports science [7-9]. However, in practice, the assessment of the Heath-Carter somatotype is not always available because of the need for a significant number of anthropometric measurements that must be performed by a qualified specialist. In our work, based on the results of anthropological study of a large group of ethnically Russian children and adolescents, we suggested simple prediction formulae for automated bioimpedance-based evaluation of the Heath-Carter somatotype that are suitable for individual typology. We could recommend preferential use of the equations based on directly measured electrical resistance rather than estimated values of fat mass index or fat-free mass index. In contrast to the results obtained earlier by the other authors [14], the formulae are suitable for use both in boys and girls in a relatively wide age range, from 7 to 17 years, and rely solely on data collected within the traditional bioimpedance measurements procedure.

Acknowledgment

The study was supported by the RFBR grants no. 13-06-00702 and 15-06-06901 (for *AVA* and *EZG*), and by the RSF grant no. 14-15-01085 (for *DVN* and *SGR*).

Conflict of Interest

The authors declare that they have no conflict of interest.

References

1. Sheldon W H, Stevens S S, Tucker W B (1940) The varieties of human physique. Harper Bros., New York
2. Stefko V G, Ostrovsky A D (1929) The schemes of clinical diagnosis of constitutional types. Biomedgiz, Moscow (in Russian)
3. Tanner J M (1964) The physique of the Olympic athlete. George Allen and Unwin, London
4. Heath B H, Carter J E L (1967) A modified somatotype method. Am J Phys Anthropol 27:57–74
5. Carter J E L, Heath B H (1990) Somatotyping – development and applications. Cambridge Univ. Press, Cambridge
6. Carter J E L (2002) The Heath-Carter anthropometric somatotype: instruction manual at http://www.somatotype.org/Heath-CarterManual.pdf
7. Kandel M, Baeyens J P, Clarys P (2014) Somatotype, training and performance in Ironman athletes. Eur J Sport Sci 14:301–308
8. Rajkumar R V (2015) Endomorphy dominance among non-athlete population in all the ranges of body mass index. Int J Physiother Res 3:1068–1074
9. Yang L-T, Wang N, Li Z-X et al. (2015) Study on the adult physique with the Heath-Carter anthropometric somatotype in the Han of Xi'an, China. Anat Sci Int DOI 10.1007/s12565-015-0283-0
10. Somatotype at http://goulding.ws/somatotype
11. Silva C A D, Mendes D D S, Oliveira E et al (2014) BodyShifter – software to determine and optimize an individual's somatotype. Procedia Technol 16:1456–1461
12. Slaughter M H, Lohman T G (1976) Relationship of body composition to somatotype. Am J Phys Anthropol 44:237–244
13. Slaughter M H, Lohman T G (1977) Relationship of body composition to somatotype in boys aged 7 to 12 years. Res Quart 48:750–758
14. Nawarycz T, Ostrowska-Nawarycz L (2001) Evaluation of the first and second components of somatotype using bioelectric impedance analysis, Proc. of XI Int. conf. on electrical bioimpedance. Oslo, Norway, 2001, pp 349–352
15. Kyle U G, Earthman C P, Pichard C, Coss-Bu J A (2015) Body composition during growth in children: limitations and perspectives of bioelectrical impedance analysis. Eur J Clin Nutr DOI 10.1038/ejcn.2015.86
16. Godina E (2011) Secular trends in some Russian populations. Anth Anz 68:367–377
17. Houtkooper L B, Going S B, Lohman T G et al. (1992) Bioelectrical impedance estimation of fat-free body mass in children and youth: a cross-validation study. J Appl Physiol 72:366–373

Address of the corresponding author:

Author: Sergey Rudnev
Institute: Institute of Numerical Mathematics
Street: Gubkin str., 8
City: Moscow
Country: Russia
Email: sergey.rudnev@gmail.com

Author Index

Keyword Index

MIX
Papier aus verantwortungsvollen Quellen
Paper from responsible sources
FSC® C105338
FSC
www.fsc.org

If you have any concerns about our products,
you can contact us on
ProductSafety@springernature.com

In case Publisher is established outside the EU,
the EU authorized representative is:
Springer Nature Customer Service Center GmbH
Europaplatz 3, 69115 Heidelberg, Germany

Printed by Libri Plureos GmbH
in Hamburg, Germany